Marimichael

MARIMICHAEL

Climate Crisis, Resistance, and the Search for
Meaning

Greg Olmsted

Dirty Business Publishing
Washington, D.C.

Library of Congress Control Number: 2020909657
ISBN: 978-1-949203-18-9
ISBN: 978-1-949203-17-2 (trade paperback)
ISBN: 978-1-949203-16-5 (E-book)

FIRST EDITION

Edited by Lori Stone Handelman, PhD
Cover design by Kerry Ellis
GregOlmstedBooks.com

DirtyBusinessPublishing.com

Dedicated to my father

Roger Orley Olmsted

Contents

ALSO BY GREG OLMSTED

STRONG CURRENT TRILOGY

UNDER WATER

UNDER GROUND

UNDER THREAT

The CURRENT AFFAIRS TRILOGY

MARITAUQUA ISLAND (WE SHALL COME AWAKE)

HAARUN BROTHERS (KLEPTOCRACY, RESISTANCE, AND THE SEARCH FOR MEANING)

OTHER NOVELS

ISTINA AND THE APOSTATE (RELIGION, GENETICS, AND THE SEARCH FOR MEANING)

1

Cosmos Club

"In summary, we must reduce CO2 emissions." Danielle Queen chopped downward through the air in front of her as if dealing a death blow to CO2. She surveyed the room, seeking eye contact with her audience of Cosmos Club members. Danielle, like her audience, was a mature intellectual, but her red face and raised voice revealed a degree of passion and anger not typically heard at Cosmos Club lectures.

Liko's date, Marimichael, had invited Liko to hear Danielle's dinner lecture at the Cosmos Club in Washington, DC. Danielle had been Marimichael's professor and mentor at West Virginia University. Liko had a seat between the two women at an eight-person circular dining table.

Liko felt his iWatch vibrate. A gray "W" on a black background appeared, the logo for a *Washington Postnewsflash*. He tactfully lowered his hand into his lap and discreetly read the message: "Congressional Websites Hacked. Hackers Demand Action. 'Vote Them Out.'"

"Divest from fossil fuel companies," Danielle said, recapturing Liko's attention. "Even the Rockefeller family withdrew all their fossil fuel holdings, and they were the heirs to an oil fortune." She targeted an elderly lady who was wearing a diamond broach. "Shouldn't we do the same?"

I don't have investments, Liko thought. *Except for my recent inheritance, I have no wealth.* He looked around the room. The members of the Cosmos Club appeared wealthy—well dressed, anyway. And smart. Membership in the club required some level of professional achievement, and the walls were lined with photos of members who had won Nobel Peace prizes and Pulitzer Prizes. Liko felt out of place.

He smiled at the members who returned his gaze. Surely Danielle was preaching to the choir.

His watch vibrated again. *The New York Times* logo, NYT, appeared and was replaced with a headline: "Oil Baron's Wife and Grandchildren Taken Hostage at New York Mansion."

What the hell?

He wanted to read the breaking news but resisted opening his iPhone, just barely. Instead, he tried to refocus on Danielle and what she was saying: "We must leave fossil fuels in the ground. We must use wind energy and solar energy instead."

Liko agreed.

Danielle then said, "I know it is unusual, but we are living in unusual times so I'm going to break with our club's normal protocol and ask one of my former students, who is also a friend, to share her thoughts about what the future holds for our fossil fuel economy. Marimichael?"

Liko glanced at Marimichael seated next to him. She smiled and he nodded encouragement. She patted his hand with confidence; did she think *he* needed reassurance that she could handle this? He realized how little he knew about her.

She rose and dropped her napkin into her chair. She was wearing a bright orange, yet otherwise tasteful, dress. He watched her walk to the podium, each step measured and confident.

At six feet, four inches tall, Marimichael towered over her petite former teacher. She hugged Danielle, took the proffered microphone, and said confidently, "Thank you Danielle." Liko had the impression that this was not the first time Danielle had invited Marimichael to speak.

Marimichael turned to the audience. "You need to divest, because the revolution is coming." She paused, perhaps waiting for a response to her mention of revolution, and then continued. "I will go a step beyond what Danielle just said: the fossil fuel economy will soon collapse because fossil fuels will be replaced by cheaper energy technologies, such as wind and solar."

Her eyes, beneath high eyebrows with strong peaks in the middle, sawed across the audience as if searching for something. Seeing startled faces throughout the audience, she broke into a cautious smile. "Do not be mistaken!" Her voice rose in intensity. "The climate change revolution has already begun. I invite you to join us." Marimichael shook her fists back and forth in front of her as if pounding on a door. "Join the revolution!"

Some members of the audience clapped, including Liko. Others sat stunned, surprised by her power and her battle cry. For the next few minutes she railed against Wells Fargo, Goldman Sachs, and BlackRock. She shamed chief financial officers by calling them out by name, including Larry Fink. *I hope none of these people are present this evening!*

She's beautiful, Liko thought, admiring her rectangular face with its broad and straight forehead and strong jaw. Her auburn hair, pulled back off her forehead, hung just above her shoulders, so thick and lush, the ends wavy and curled.

"Join our revolution," Marimichael repeated. She handed the microphone back to Danielle and returned to her table to scattered applause.

Liko stood up quickly. He was the same height as Marimichael. He nodded and said, "Well done" as he pulled out her chair and guided her to her seat like a perfect gentleman.

When the question-and-answer period ended, and after they said goodbye to the folks at their table and to Danielle, Liko found himself standing next to Marimichael at the front entrance of the club. They faced Massachusetts Avenue on the edge of a semi-circular driveway in front of the heavy Beaux Arts façade.

During the lecture, the temperature had plummeted from the mid-fifties to seventeen degrees as a record-breaking arctic blast descended on DC. Snow flurries swirled in the air but melted as soon as they touched the ground.

A black SUV pulled up. A man in the passenger seat jumped out, opened the back door, and stood expectantly.

Marimichael turned to Liko, gave him an unexpected, bracing hug, and placed something into his hand. She stepped abruptly into the back seat of the SUV and then she was gone. Not even a "goodbye," or a "see you later." But she did say, yet again, "Join us."

Shocked, Liko looked down at the object in his hand. It was a thumb drive with a bamboo case and an owl icon printed on both sides. He placed it into his front pants pocket.

His watch vibrated. A weather app announced that a hypothermia alert had been issued and emergency hypothermia shelters had opened for the homeless in DC. He shivered, involuntarily, and rubbed his hands together. It felt more like January than the day before Veterans Day, which was ordinarily brisk but sunny.

As he walked the five blocks from the club to his condominium, Liko felt that everything in the world had changed. He heard fire engines in the distance. In the direction of the White House he heard police and ambulance sirens.

His watch vibrated again and he read the *Washington Post* mini-newsflash: "Danson Assassinated at White House Ceremony. Suspect Killed by Old Guard."

He pulled the collar of his dress coat up around his neck and stuffed his hands deep into his coat pockets.

2

Loneliness

Liko awoke the next morning, Veterans Day, alone in his king-sized bed. The white bedsheets were on the floor and he was naked. He hated pajamas. As he swung his feet over the side of the bed, the dramatic events of the night before came crashing into his awareness: A soldier had assassinated the President of the United States. Activists had hacked into congressional websites. Someone had kidnapped the wife and grandchildren of an oil baron. And Marimichael had jumped into a black SUV, abandoning him on the front steps of the Cosmos Club.

Liko stood up, donned his flannel robe, which he always kept within reach of the bed, and shuffled to his bedroom door and pushed it open. He slept naked, but he was too modest to walk around naked.

He walked to the bank of windows in his living room, with a sixth-floor view, and gazed outside. Directly across the street were three hotels. The monotony of their façades was broken by the changing color of brick: burnt orange, red, and then dull yellow. In contrast, his new condominium building was a modern design of gray metal framing and large glass windows.

He walked to his kitchen and fixed a cup of hot water with lemon juice and honey and then sat down at his dining table. He would have preferred to sit outside on the porch, but that would have required him to walk through his guest's bedroom, and she was probably still asleep. He couldn't fault her for sleeping in; it was still early morning.

Where did Marimichael go? he wondered. *What was that black SUV about?* He ran his index finger along his eyebrow and pulled his ear lobe. *Why did she leave so abruptly?*

He had met Marimichael for the first time at the Dupont Circle farmers market the previous Sunday, seven days ago, just before noon. He could not stop thinking about her. It had been a beautiful blue-sky day with sunshine dominating and a temperature trying to reach seventy, a warm day for November.

Marimichael had stopped abruptly in front of a kombucha vendor to greet a mother pushing a stroller with an infant. She had stopped so suddenly that Liko almost ran into her. That's when he noticed her tattoo. On her upper back, the low halter top revealed a meteor flashing across the sky, from shoulder to shoulder. Glimpsing the meteor was like seeing it flare across the sky.

Although she had been the one to stop suddenly, Liko had found himself apologizing profusely. And then he had found himself tongue-tied, so he quietly walked on.

After passing several vegetable vendors, he had stopped to sample yellow and white peaches, nectarines, and plums that had been sliced and set out on white paper plates for prospective customers to taste. Once she walked by, he had followed her.

He couldn't help himself. He wanted to talk to her, but he was shy. He looked her over from a distance. She was beautiful. He moved closer. On her right shoulder, hatched in black ink, a woman sharpshooter kneeled, posed to fire. On her left shoulder, a black vine snaked and blue flowers bloomed. Liko could not identify the species.

He watched her interact with others—such charisma. She had a way of engaging people, blocking out all distractions and making time stop, even though the farmers market was bustling with activity. Every person she talked to laughed, smiled, and become totally engaged with her. She had mesmerized him, too, as if casting a spell.

And then she noticed him. "Aren't you the guy who gave that speech at Davos?"

"Yes." *How did she know? She's even more interesting than I imagined,* Liko thought.

"That was wonderful!" She looked him up and down, taking the measure of him. "You knocked out Derichenko, one of the most powerful oil titans in the world. And the most heavily guarded. Impressive!"

"Yes," Liko admitted. "But I didn't know who he was when I knocked him out. I just thought he was an arrogant, rich jerk."

"But still," she said.

"He has threatened me," Liko said, surprised at how easily she got him to reveal a dark threat that now hung over his life.

She looked at him carefully, again taking the measure of him. "But he doesn't frighten you?"

"No." Liko answered honestly. "But if he comes for me... Well, I might not see him coming."

She nodded her head, and he wondered if she understood.

She walked ahead of him, just a few steps, stopping to look at fruits or vegetables or crafts, allowing him to catch up. They would say a few words then continue their walk, him following close behind, almost with her, almost together, exploring with their senses wide awake. Liko felt a motivation—a need—that he didn't understand.

It was an animal-like courtship. Uncalculated. On the level of pheromones and hormones, and smells and tastes and the unexpected texture of fuzzy peaches, and the sight of plum juice on her lips and chin. He wiped her chin with his fingertips and the juice was sticky and sweet.

A vendor, a young girl, allowed them to share a pluot. It was the first pluot that Liko had tasted, and the woman with the meteor tattoo shared it with him. They smiled together. Together they looked at beeswax candles and then sampled different honeys: cranberry, cherry, and wildflower.

"Would you like to have dinner this evening?"

"Oh, yes," Liko answered.

She wrote her address on the back of the receipt for tomatillos that he had bought her.

"What time?" he asked.

"Seven." She smiled. "My name is Marimichael."

"Liko," he said, placing his hand on his chest.

3

Shane

That evening, last Sunday, at exactly 7pm, Liko had knocked softly on Marimichael's front door. Her apartment was a two-story, historic brick townhouse, centrally located in Georgetown. Heavy coats of white paint covered the brick façade. Windblown oak leaves rested along a black iron fence, unraked.

The front door opened. "I'm so glad you could make it!" Marimichael's face filled with a smile. Her teeth were large and white.

"Me too," Liko said. He returned her smile.

His naïve smile was genuine and heartfelt, and that seemed to please her. "The sunflower is beautiful!" With her free hand she reached out and took from him the cheerful, bright yellow flower. She held it between them for a moment. She started to say something, but then seemed to change her mind, saying instead, "It's huge! Thank you!"

Liko's smile widened.

She stared directly into his brown eyes for one, two, three seconds and then she turned her head modestly to the side and looked at the sunflower. When she looked up, he was still smiling. "And wine, too?" she asked.

He nodded yes, without actually saying yes. Just a gentle nod of his head.

She shifted her weight to keep the door open behind her, and then she took the bottle of wine in the brown paper sack from him. The paper ruffled in her hands. "Come on in."

He stepped into the foyer. The floors were hardwood and the room had ten-foot ceilings and detailed molding. She directed him to the dining room.

Liko saw the bowl of green salad on the dining table: shredded carrot, sliced red onion, and bright red tomatoes, hacked into large chunks, all resting in a bed of greens.

"From the farmers market?"

"Of course," she answered.

A floor lamp in the adjoining room drew his attention. The light shone down on a man sitting in the high-backed red chair, surrounding him in a yellowish-white spotlight.

"Liko, I'd like you to meet my brother."

Liko glanced from Marimichael to the man and then back to Marimichael. This time she did not meet his gaze and he found no place to rest his eyes, so he looked back at her brother.

The man sprang from his seat. He crossed the room and extended his hand.

Liko took a step backwards, reflexively, into a boxer's stance.

"I'm Shane," the man said. He was wearing jeans and a

red and black plaid shirt, tightly tucked, with the sleeves rolled up.

"I'm Liko." He shook Shane's hand. He had a strong, confident grip.

Liko noted that Shane was several inches taller than he was—which rarely happened. He was also broader in the chest and narrower in the waist, with larger forearms and biceps: the physique of someone in excellent health. Liko, being Hawaiian and a bodybuilder, wasn't used to meeting someone larger and stronger than himself.

"Dinner is almost ready," Marimichael said. "I just need to check the ribs."

Liko's broad nostrils flared and he smelled barbecue.

Marimichael disappeared into the kitchen with the sunflower and the bottle of Trader Joe's Cabernet. "You guys get acquainted," she said, raising her voice. "I'll finish up in here."

Shane led Liko into the den and directed him to another high-backed wing chair. Also red.

Shane turned off the warm spotlight, leaving only the single incandescent bulb hanging in the center of a light fixture above the dining room table and the glow from the television set in front of them.

Liko recognized the movie playing. It was *First Reform* with Ethan Hawke. He recognized the funeral scene; it was filmed at Arthur Kills, Stanton Island, among scrap metal, old ships, and sediments filled with toxic metals. The music was classic Neil Young. A very pregnant woman was scattering the ashes of her husband, an unstable environmental activist, who had committed suicide. Ethan Hawke, dressed in the black robe of a priest, was officiating the unorthodox funeral.

"I love this movie," Shane said.

"I saw it in the theater," Liko said, "on the big screen."

"The director also did *Taxi Driver*."

"And *Raging Bull*," Liko added, "one of *my* favorites." He thought Robert De Niro was excellent in the role of Jake LaMotta, the violent boxer who spiraled into self-destruction. *I need to get back into boxing*, Liko thought.

Liko felt that Shane was looking at him. He tried to keep his eyes on the television, though.

Will he approve of me seeing his sister? And if he doesn't?

4

Dinner
Discussions

When Marimichael called them to dinner, Liko saw that she had placed his sunflower in a tall, cylindrical vase in the center of the table, directly beneath the overhead light. That pleased him. The yellow flower head, pregnant with large black seeds, had bowed in the direction of her seat, as if in reverence.

She had also placed a rack of ribs on each of their plates and a roll of paper towels between Shane and himself. She passed Liko a plate heaped with steaming white corn.

Liko took the plate from her and picked up an ear of corn. Hot! Even the plate, which he quickly set down between himself and Shane, was hot. Liko tossed the corn hand to hand until he dropped it onto his meaty ribs, which were slathered in a dark barbecue sauce. He suddenly realized how hungry he was.

He glanced at Marimichael and felt his cheeks burning. She smiled and laughed, and so did Shane. All three of them laughed together.

They ate quietly, tackling their salads first, until Marimichael broke the silence. "Mankind is causing a mass extinction: rhinos, polar bears, and even emperor penguins." She looked at Liko. "It is so sad."

Liko had just forked lettuce and other greens into his mouth. *I think you like animals,* he thought, *and babies in strollers at the farmers market.* He chewed quickly and, after swallowing said, "Yeah. It's terrible."

"How can people be so complacent?" she asked. "People need to stand up, take back their government, demand action."

"I agree," Liko said.

"Must there be a crisis before people wake up?" she asked.

Liko didn't know what to say to that, and wished they could talk about something else.

"Like a big chunk of Greenland breaking off?" Shane suggested.

"Or the Antarctic Ice Sheet," Marimichael added.

"Let's hope for a major breakthrough in technology," Liko said as he slathered the white corn with real butter and a light sprinkling of salt. The butter melted and flowed onto the ribs.

"But we already have solar and wind," Shane countered. "The problem is not a lack of technology, but the lack of a leader. We need a leader people will follow, and who will motivate us."

"I read that local and state governments are taking the lead," Liko said. "I mean, because Trump pulled out of the Paris Agreement."

"Trump!?" Marimichael rolled her eyes and shook her head. "Thank god he is gone, finally. But we still need bolder action."

"California is leading the way," Liko said. "They reduced air pollution in Los Angeles. They have a lot of good ideas about how to do things."

"We still need national leadership," Shane said.

Liko discovered ceramic corn holders next to his fork. He speared the buttered corn and set it on the side of his plate, next to the ribs.

Marimichael passed Liko the salad and he refilled his salad bowl. He recognized the fresh vegetables they had chosen together at the farmers market: red onion, Bibb lettuce, and dark heirloom tomatoes.

"What about young people?" Liko suggested. "They're motivated, and they have a lot of energy."

"But they cannot fix the problem." Marimichael said. "They are good at social media, but they only respond to each other. They don't communicate with their parents or other adults. Just their peers."

"They attend protests," Liko countered. "And they've been walking out of their schools."

"So? What has that accomplished?" Shane asked Liko.

"What about the Swedish girl," Liko said, "What's her name?"

"Greta Thunberg?" Marimichael said.

"Yes," Liko said, "Greta sailed across the Atlantic in a solar- and wind-powered boat."

"Correction," Marimichael said, "she was a *passenger* in a solar- and wind-powered boat. Adults did the sailing. And it will be adults who end global warming. If we wait for the children to grow up, get educated, and learn about the world and government, it will be too late."

"Liko, young people have little knowledge of life," Shane added, agreeing with his sister.

"And you can't get into the school curriculum to influence them because it is so closely guarded," Marimichael said.

"There is no way to get in front of the kids," Shane said, once again agreeing with his sister.

"Parents abandon their kids to popular culture," Marimichael said. "And when they are ten, eleven, and twelve years old, parents and teachers don't provide leadership."

"And by the sixth grade sexuality kicks in." Shane smiled at Liko. "Remember the sixth grade, Liko?"

Remembering his sixth-grade experience, Liko frowned. He had grown up in a trailer park outside Las Vegas with his alcoholic mother. When he was in the sixth grade, his thoughts were about running away from home. Not girls.

"By the time they are teenagers, they have been taught that their parents are the enemy," Shane added.

The enemy? Again Liko frowned. Yes, his father actually *was* the enemy—an alcoholic gambler who abused Liko's mother. But she had left him, and Liko always admired her for that. She had protected him.

"Like Shane, I have no hope in young people solving the climate crisis." Marimichael had set the rib bone down on her plate and was sucking the sweet barbecue sauce off her thumb and forefingers. "There is just not enough time. They are uneducated and know nothing about politics. We need action now."

"I disagree," Liko said. "I think young people *are* taking action."

"Like what?" Marimichael and Shane asked in unison.

Liko was beginning to feel uncomfortable. *Why so serious? And why gang up on me?* He almost joked about the meaty ribs. If they were such die-hard environmentalists, he thought, why did they eat meat, and especially beef?

"Some are getting arrested and taken to court," he said. "They formed the Sunrise Movement. Some, like Greta, have been trolled in person and online."

"But what have they changed?" Shane asked.

"They've brought attention to the crisis," he said. "It wakes up parents and grandparents. It educates the public."

"Does it?" Shane asked. "This is an urgent crisis and we have to move fast." He picked up an ear of corn off the plate in front of him without using ceramic corn holders. "I do not see how being arrested accomplishes anything."

Liko noted that the corn must have cooled because Shane handled it without problem. Unlike Liko, he ate it without butter or salt.

"But again, what about the Sunrise Movement?" Liko asked. "They've done things, including sit-ins and protests and promoting the Green New Deal."

Shane gnawed one side of the corn. "Once those kids get their inheritance, their political views will change."

"How can you know that?" Liko asked, once again disagreeing. "I think we should stand in solidarity with young activists. After all, it's their future we're trying to save. And hopefully one that's livable."

Shane turned the corn to expose an uneaten row of kernels. He stared across the table intently at Liko, his eyes brown and focused, sharply.

Uh-oh! Liko thought. *I didn't mean to upset him.* He tried to think of something positive. "Now that we have a Democratic president, won't things change?"

"Not as long as Senator Nappe controls the Senate and uses the filibuster."

Senator Nappe? Who's he? Liko wanted to ask but he didn't want to appear ignorant. He would Google Senator Nappe later. "If he's a roadblock, then vote him out."

Marimichael laughed. "But he was just elected for another six years!"

Liko noted that Marimichael was also looking at him. Perhaps judging him. He sensed that he was disappointing her, too. *But why?* He shifted uneasily in his chair.

"The system is not working for the people," Shane said. "Someone has to fix the system."

"Or change it," Marimichael said. "And now."

"You don't think democracy's working?" Liko asked.

"Do you?" Shane asked.

But before Liko could answer, Marimichael said, "Dark money—especially fossil fuel money—has corrupted American democracy. Fossil fuel companies and a group of anti-regulatory billionaires have *bought* Congress."

"The fossil fuel companies have a lot of money and a lot of power," Shane added. "They have lied to the public and deliberately confused people. They even deny that climate change exists."

"Their think tanks have turned many Americans against science," Marimichael said. "Isn't it incredible? That they have been able to do that is shocking."

"They are interested in profits," Shane added, "not the future of our planet."

"And imagine this," Marimichael said, "When Trump was president, he appointed a Secretary of State who was an ExxonMobil chairman, and who received Russia's "Order of Friendship" from Vladimir Putin!"

"So what's the answer?" Liko asked.

"Protest," Shane answered.

"Like the seventies," Marimichael added.

Shane finished the corn and placed the cob on his plate next to a stack of rib bones. "People marched in the street then. They demanded change. They were all over Capitol Hill. It was raw political power."

"It all starts with political power," Marimichael said.

So what does any of this have to do with me? It's all out of my control, Liko thought. *And you guys need to lighten up or I'm out of here.*

5

Tattoos

"Next Sunday," Marimichael said, "a friend of mine, an investigative reporter, is presenting at the Cosmos Club. She will be discussing divestiture from fossil fuel companies. Are you interested?"

"Sure," Liko answered.

"Great! It's a dinner presentation, so you'll need a coat and tie."

Liko nodded. He had a Canali suit, a custom-tailored suit he had bought in Italy. "No problem. I'd like to go."

"Her name is Danielle. I took her journalism class when I was an undergraduate. She is an environmental journalist."

"An environmental journalist? I didn't know there was such a thing."

"Oh, yes. There is even a Society of Environmental Journalists." Marimichael set her hand on top of his. She then got up from the table and carried her plate and Liko's into the kitchen. He tried to help clear the table, but she insisted he stay seated, and visit with Shane.

"Where do you live?" Shane asked as Marimichael left the room.

"I have a condominium nearby," Liko answered.

"Are you working?"

"No, not right now. I used to work as a security guard."

"I saw the video of you at Davos. That was awesome."

"Davos? Yeah, we were demonstrating against a kleptocrat named Lil't."

"It was amazing. Especially how you used the drones."

Liko recalled how the drones had chased Lil't off the stage. They had streamed confetti on the dinner audience and then sprayed paint over a mural of Little Neom. "I can't take credit for that. Others operated the drones."

"But you guys started a revolution! You brought down a kleptocracy, the Republican Democracy."

"I did what I could."

Shane fired question after question, and Liko patiently answered.

To Liko's dismay, Shane then changed the conversation back to the climate crisis. "America is paralyzed. We need to break the gridlock, end the paralysis, and start a revolution."

Damn! Liko thought. *That's not what I want to talk about.*

Liko decided that he had heard enough about the climate crisis for one visit. Besides, there was nothing, absolutely nothing he could do that would lead to positive action. It was all so depressing. Why talk about it, when there was nothing to do?

"We need a military dictatorship," Shane stated, matter-of-factly. "Like China."

Liko frowned. "You don't really mean a dictatorship, do you?"

"Would it be better if humans and all the other mammals disappeared?"

Of course not, Liko thought. But he did not want a dictatorship, either. "I believe in our democracy."

Marimichael had returned from the kitchen and was standing beside Shane's chair. Her hand rested on his shoulder. Brother and sister. Both tall. Both strong and determined.

Liko watched Shane looking up into his sister's eyes; without saying a word, he seemed to be conveying his disappointment. It was obvious to Liko that he was failing some test and disappointing them both.

"Well then," Marimichael said. "You need to come up with a way to fix American democracy: the gerrymandering, the misuse of the Senate filibuster, the influence of corporate money on politics. And you need to fix it *yesterday.*"

Liko had no solution. He had never seriously considered how to address any of those problems other than by voting.

"Americans have allowed corporations to take over their government," Shane said. "Perhaps now they will allow the military?"

Impossible, Liko thought. *That will never happen.* "Americans have too many guns. They would resist."

Shane smiled. He leaned forward with his elbow to the side of his plate, the knuckle of his index finger covering his lower lip. "Americans with guns will lead the revolution."

"Whoa!" Liko said. He raised his hands with his palms facing Shane. "I really can't see that happening."

Shane again looked up at Marimichael. Their eyes met and they held each other's gaze for a moment. As if answering an unasked question, Marimichael raised an eyebrow and shrugged her shoulders, ever so slightly.

Liko wondered what they had communicated to each other. "Thank you for dinner, Marimichael. I enjoyed it."

"You're welcome." She smiled. "Let me show you the house?"

"I'd like that," he said.

She offered Liko her hand and he accepted it.

The second floor had a large master bedroom suite with an attached office. The smaller bedroom, with the red roses and calla lily wallpaper, was hers.

Being in such an intimate place, Liko dared to ask her about her tattoos.

She surprised him and stripped off her shirt. Turning her back to him she said, "You can unhook my bra so you can see the meteor better."

He unfastened it. She slipped out of the bra and set it on the brown comforter that loosely covered the bed in front of her.

Her back was broad and muscular. He massaged her neck muscles, briefly, as he admired her upper back muscles and the meteor tattoo.

"The comet is magnificent," he said.

"It's a meteor," she said. "Not a comet."

"What's the difference?"

She turned her head and looked over her shoulder at him. "It's the Great Meteor of 1860. But I don't think you're ready to hear about it."

He placed his hands on her shoulders and felt her powerful deltoids: the rounded lateral, and the chiseled front and rear deltoids. He squeezed her shoulders and upper body between his hands, firmly, as if hugging her. His pec muscles flexed. She was solid, powerful. "You have the build of a swimmer."

"Fifty laps a day."

"I don't like chlorine," Liko said. "I prefer the ocean."

"I've never swum in the ocean."

"Never?"

"Never."

He turned her a quarter turn so he could look at the woman warrior on her shoulder, kneeling and aiming a rifle.

"It's a sharpshooter tattoo," she said, before he could ask the question. "I earned it during my ten years of service in the army."

She turned her other shoulder to him. "Wild blue lupine flowers and an endangered Karner blue butterfly. I worked to save them."

He ran his finger along the lower petals of the lupine flower, outlining its cup shape. He then turned her to face him. He pulled her to him and held her in his arms for a few moments.

"Don't worry about him," Marimichael said, guessing his thoughts. "He knows I like you."

Liko backed her up against the edge of the mattress. With her brother downstairs, it felt dangerous. Liko found that exciting.

"Does your bed squeak?" It was a queen size sleigh bed.

She wrapped one of her legs around him and he pushed her gently onto the comforter. She ran her fingertips over his thick black eyebrows, flat and broad nose, and heavy lips. "You are handsome."

"Is that why you like me?"

"I love sex," she answered.

6

Jack, an FBI
Man

Liko took the last sip of hot lemon and honey water and
set the mug on his kitchen table. As the mug struck the
glass surface, it jarred him back to the present, and all
the tragic events of the previous night came flooding back
into his mind: Congressional websites hacked, the wife of
an oil baron kidnapped, and the President of the United
States assassinated. *What the hell is happening?*

And why had Marimichael left so abruptly in the black
Cadillac Escalade SUV after the lecture, leaving him
standing alone in front of the club? That had embarrassed
him, even though she had given him a quick hug before
leaving.

And what was on the thumb drive she had slipped into
his hand? And where was it? He glanced at the granite
kitchen island and counters, the tabletop, and the coffee

table in the living area, but he did not see it. Maybe he had left it in his suit coat pocket?

He got up from the kitchen table and walked back into the long hallway that led to the master bathroom. Cabinet doors lined both sides of the hallway, opening to wardrobe storage, either shelves or hangers.

He still needed to clean out the previous owner's clothes. Zahi had been tortured to death in the Republican Democracy, and in his will left Liko not only his condominium, but also his possessions, including his silver Lexus RX350, and a considerable amount of money in bank accounts and investments. Liko had put off the cleaning because going through Zahi's personal possessions felt awkward.

Liko opened several more doors before he found his dark gray suit. He took out his suit jacket and checked the breast pocket. The thumb drive wasn't there. He checked the opposite pocket. Again, it wasn't there.

He hung the suit jacket in the closet and took out the dark gray pants. There it was in the front pocket: a thumb drive with a bamboo case, stamped with an owl icon. He hung the pants back up.

He returned to the living room and set the thumb drive on the coffee table. *I'll look at it after my shower,* he decided.

A loud buzzer startled him; he waited until he heard the jarring noise again and located the source: a panel on the wall. He walked over and discovered a security system with a liquid crystal display panel, light emitting diodes, and a plethora of mysterious buttons.

The loud buzzer sounded again. Frustrated, trying to read and understand the meanings of all the buttons, he randomly pushed one after another, while saying "Hello? Hello? Hello?"

A voice said, "Mr. Koholua, a man from the FBI is here to see you. Should I send him up?"

When Liko answered the door, he thought there had been a mistake. The man standing in front of him didn't look like an FBI man. He was a short black man wearing a red straw hat. *Perhaps to make himself look taller,* Liko thought. His black pants stretched tight over his legs, his black coat was cut too short and decorated with colorful embroidery on the lapels and shoulders and back, and his shoes were similar to cowboy boots, yet not quite.

"Do you have ID?" Liko asked. Liko was twice as hefty as the slightly built man. He stood, firmly planted in the doorway, to prevent the man from entering.

The man took off his hat and held it in his left hand with a newspaper. His hair was perfectly coifed. In one smooth motion, the man took out a slim leather wallet from his pants pocket and flipped it open, revealing a picture ID on one side and a shiny gold badge on the other. A gold eagle rested at the top of the badge and Liko read "Federal Bureau of Investigation, US" and "Special Investigator." The picture on the ID matched the man's face: round with large lips. The thought *juicy lips* flashed into Liko's mind and he wondered if that was racist. He hoped not. The picture also showed a gypsy nose ring, just like the one dangling in the nose of the man standing in front of him. The name on the ID read Jack Amaya.

"My name is Jack Amaya. I work for the Federal Bureau of Investigation."

Jack Amaya? You've got to be kidding?

"I'm a forensic accounting investigator. I direct the Bureau's Public Corruption program in DC. Are you Liko Koholua?"

Damn, Liko thought, remembering his recent inheritance. "Yes," he answered. "Come on in." He held the door for the FBI man.

The man entered.

"This must be about Zahi's estate?" Liko asked. "Do I owe taxes, or something?"

The man's expression was quizzical, puzzled. One eyebrow raised. "Not that I am aware of."

"Really?" Liko pursed his lips in surprise. "Have a seat." He directed the FBI man towards the living room, allowing him to choose one of the white cushioned armchairs or the white couch.

Jack sat down in an armchair and dropped his red straw hat to the floor beside him. He set the newspaper in his lap. Liko sat across from him in a gray B&B Italia chair. They were separated by the length of the coffee table. Bright morning light filtered through the sheer curtains.

Jack's eyes followed the rays of the sun to the base of the lamp on the table next to him. "An abacus?"

"Yes. The previous owner was an economist," Liko said. "The lamp might be custom made, I'm not sure."

Jack reached out and moved the beads on the abacus. "It is functional." There was surprise in his voice. "Nineteen plus seven," he slid the beads back and forth along their guides, "is twenty-six." Satisfied, he leaned back in his seat.

"I'm sorry to visit at such an early hour." Jack raised his hands in front of his chest and then swept them from his shoulders to his knees, highlighting his outfit. "I was at a dance last night when the news broke—all the terrible news. I got called back to work and haven't had a chance to change." He smiled and said, "I'm a flamenco dancer."

"Oh. I see." Liko nodded. He suddenly realized that he

was still wearing his flannel robe. He pulled it closed in front and leaned back in his chair. "No problem. What's the purpose of your visit?"

Jack picked up the newspaper from his lap. "Your newspaper?" He held it out towards Liko. "It was in the hallway, on the door handle."

Liko rose to his feet, took the newspaper from Jack, and set it on the coffee table between them. It unfolded, opening to the front page. It was the *Wall Street Journal*—Zahi's annual subscription was still active—and the headline screamed "President Danson Assassinated!"

Liko was taken aback. "I heard about that last night. On my way home."

"It's a tragedy," the agent said.

"Yes, it's terrible."

"Do you know a man named Shane O'Brien?" He removed a picture from the inside pocket of his short jacket and held it out to Liko.

Liko leaned forward to look at the picture in Jack's outstretched hand. He recognized Shane. It appeared to be a recent picture; who could mistake his curly brown hair?

"Yes. I met him a few nights ago. Last Sunday night. The first and last time I saw him."

"Where did you meet him?"

"At his home. On P street. I walked there from here. It's close by."

"What was the purpose of your visit?"

"Dinner."

Jack wanted to know how he had met Shane and why he had dinner with him. Liko explained that he had met Jack's sister, Marimichael, at the farmers market around noon, and that she had invited Liko to dinner that evening.

"She invited you to dinner at her home in Georgetown?"

"Yes."

"Who else was there?"

"No one. Just the three of us: me, Marimichael, and her brother Shane."

"What did you discuss?"

Liko stared at Jack and held his breath a moment. He felt a tingling, an uneasiness. "Did something happen to Shane or his sister? Am I being charged with a crime?"

At that moment, Charla stepped out of the guest bedroom. She was mostly dressed, but not completely. It was obvious she was a man who was a woman, in a short, sheer nightgown that revealed her female breasts, and navy blue speedos that had a distinctive male bulge.

Jack's jaw dropped and he looked from Charla to Liko, and then back to Charla. "No. You are not being charged with a crime."

"Say what?" Charla said. She put her hand on her hip and stared back at Jack. Her mouth pouted and her eyebrows frowned at him. Her hair was disheveled.

"Have you seen Shane since last Sunday?" Jack meant the question for Liko, but he was still staring at Charla.

"Who is Shane?" Charla asked.

Jack abruptly turned his attention back to Liko, effectively dismissing Charla. "Shane was involved in a kidnapping last night."

"Really?" Liko and Charla said in unison.

Jack looked back at Charla and said, "Do you mind, Mr.... Ms.... whatever? I am here on an official visit. This is a private conversation."

"Don't mind me," Charla said. She walked gracefully

into the kitchen, opened the upper refrigerator door, and looked inside.

Liko knew Charla well enough to suspect that she was hiding behind the open refrigerator door so she could listen. *Good grief.*

"He broke into Augustus Dimer's mansion last night. He has taken Augustus Dimer's wife and two grandchildren hostage."

"Oh my god!" Charla said. "Dimer? The billionaire?"

"Do you know Mr. Shane O'Brien?" Jack asked, turning his upper body to face Charla.

"I don't know kidnappers," Charla said. She closed the refrigerator door hard enough for the contents on the door shelves to rattle. She then pranced back into her bedroom, leaving the door ajar.

"My guest," Liko said, taking a deep breath. He didn't want to discuss Charla. He didn't want to explain that she was visiting for a week or two or three. "She's a friend."

Jack looked at Liko and said nothing.

"She's a good friend. She wasn't at the dinner and she hasn't met Shane or his sister."

"I see," Jack said. "The reason I'm here is because the kidnapper—Shane O'Brien—he has requested you serve as an intermediary between himself and the FBI."

Liko felt his heart jump sideways. "An intermediary?"

"Yes. A go-between."

"But I don't want to be involved."

Jack said nothing, yet he continued to look at Liko, now more closely.

Liko repeated, "I don't want to be involved."

"He specifically asked for you."

"I don't know why," Liko said. *Unless he was impressed with what I did in Davos, but that was just smoke and mirrors.*

I'm out of my depth, here. Liko said slowly and as clearly as possible, "I ... do... NOT... want... to be... involved."

Jack shook his head, disappointed. He slapped his thigh and seemed to refocus.

Liko wondered if Jack's tightly fitting nose ring was real. He thought it might be.

Jack gave Liko his card, then picked up his straw hat off the floor and left.

"I'll take that." Charla stepped from her bedroom doorway, crossed the room, and snatched the business card out of Liko's hand.

Liko sat in his Italia chair and stared at her as she read the card with intense interest.

7

Charla

"You need pajamas," Liko told Charla.

"We both do."

Liko owned no pajamas. "At least my robe is flannel," he tried to smile, "and not see-through."

Charla adjusted her sheer nightgown. She moved to the chair that the FBI man, Jack, had sat in minutes before. She crossed her long, naked legs. "I have a kimono."

"That might work." Liko tried to imagine her in a bright kimono. "Is it short or long?"

"Liko, you surprise me!"

Now he did smile. He knew that very little surprised Charla, even though she often acted surprised.

"Wasn't he cute?" she said. "In his tight black pants and short vest?"

"And red straw hat?"

"I can imagine him tapping around that hat."

They both laughed.

"But seriously," Liko said. "Can you believe this shit? I meet a girl at the farmers market. She invites me to dinner. I meet her brother, who later in the week kidnaps the wife of a billionaire?"

"Yeah, that's crazy."

Liko leaned back in his gray chair.

Charla uncrossed and recrossed her long legs. "Did you tell the FBI man about the lecture?"

"What do you mean?" The question annoyed Liko.

"You had a second date with her, didn't you? At the Cosmos Club? Something about investments, right?"

"I saw her, yes. But it was not a date. And it was *di*vestments, not *in*vestments."

Liko turned on CNN to change the subject.

"Not since President Eisenhower has a retired army general served in the White House," a CNN commentator said. "Until today."

On screen appeared a picture of Vice President Charles Clincher, one hand on a bible, his right hand raised. The man standing next to him looked familiar, but Liko could not place him. Liko looked at Charla just as she turned her head to look at him. They both shook their heads in dismay at what they were watching.

The CNN commentator continued: "President Clincher took the oath of office last night in the National Military Command Center."

A second commentator asked, "Has the White House explained why the Pentagon's War Room was chosen as the location for the presidential oath of office?"

The first commentator answered, "No, but the message to our adversaries is clear. President Clincher is now Commander in Chief. The United States military is locked and loaded. Clincher is ready to defend our country, our life, and our liberty."

The picture on the screen did a Ken Burns zoom-in on the man in the photo standing next to the president.

The first commentator continued: "According to a high-level source in the White House, the new president insisted that the Speaker of the House, Mr. Stewart, give him the presidential oath of office instead of the Chief Justice of the United States Supreme Court. According to the source, members of the Court were upset. Nevertheless, President Clincher wanted the public to know that he was the leader of the American people and that he had the Speaker's support."

The second commentator said, "Yes, I would agree with that assessment."

"Good grief!" Charla said. "That's all we need, a general in charge of our country."

"Let me see what I can find." Liko Googled "Clincher biography" on his phone and then selected a page. "His vice-presidential biography is still posted. Interested?"

Yes, Charla nodded. She muted the television volume, set the remote control on the coffee table, and gave Liko her attention.

"According to the White House web page, Clincher first attended West Virginia University but later graduated from West Point, in New York." Liko skimmed several paragraphs. "He participated in the invasion of Afghanistan during Operation Enduring Freedom and in the invasion of Iraq during Operation Iraqi Freedom." Liko scrolled through the list of awards and medals he had received.

"In an interview he said 'Climate change is a threat multiplier.'"

"That's good," Charla said.

"After he retired from the military he worked as a cyber-security expert and supported the defense department." Liko paused to allow Charla to comment. When she didn't, he continued. "Clincher has a long list of certifications—CISSP, ISSMP—and many more. And there is a lot more about social media warfare and cyberattacks and drones."

"Cybersecurity expert?" Charla asked.

"Yep."

Liko stood up. "I need to de-stress. I'm going to the gym across the street to lift weights."

When Liko returned from the gym, he found Charla's bedroom door closed. He thought she had gone out or that she didn't want to be disturbed, so he took a quick shower, changed into jeans and a black Nike T-shirt, and returned to the living room.

He inserted Marimichael's thumb drive into his old MacBook Pro and turned on Apple TV. The thumb drive icon appeared on the television screen as "Climate Crisis Materials." He opened the drive and the first folder, which was labeled "PHOTOS, VIDEOS, AND REFERENCES."

The first photo showed a young Marimichael, maybe twelve years of age, seated in a roughly built treehouse surrounded by a group of young boys. Marimichael wore blue coveralls and was barefoot. She sat with her legs and feet dangling. Beside her sat a boy who Liko recognized as Shane. He was also wearing coveralls.

Standing behind Marimichael in the treehouse was another boy, and he too was wearing coveralls, the same kind as Shane and Marimichael. Another brother? The facial features looked similar.

Below them on the ground stood a gangly kid with a facial birthmark. Poor kid, Liko thought. The birthmark covered part of the boy's cheek and the corner of his mouth.

Next Liko clicked a video. It showed school kids protesting in front of the White House for action on climate change. The front row of kids held a banner that read "CLIMATE ACTION NOW!" They were chanting, "What do we want? Climate action!" He closed the video.

Liko opened one of the Word documents. It listed climate change references: movies, books, reports, and webpages. Liko surmised they were meant for him to study.

Next he looked inside the folder "BOUGHT AND SOLD." He opened one titled "Dark Money Influence in Congress." He perused several newspaper articles. One suggested that electric companies in Virginia controlled the Virginia State Legislature and, therefore, got whatever they wanted.

Inside the last folder, "CLIMATE CHANGE REVOLUTION," was a pamphlet: "The Legacy of John Brown and Climate Change Revolution, by Marimichael O'Brien," and a Word document titled "ForYou.doc."

He opened the document. The message was short and simple: "We will be contacting you soon. Join us." There was no signature, but Liko knew it was from Marimichael. It was her trademark phrase.

Charla came out of the bedroom and lounged on the couch in a kimono. She looked at the television and read Marimichael's invitation to Liko asking him to join the revolution. Charla rolled her eyes. "Who is this woman?"

"I don't know," Liko confessed. "I really don't know."

"She emailed you?"

"No," Liko said. "It's on the thumb drive she slipped it into my hand after the lecture. Before she jumped into the SUV and sped away."

"Shit! Liko, you need to tell the FBI."

"Why?"

"Why? Charla raised her voice. "The woman is crazy, Liko! Look what her brother did. He kidnapped children. And she wants you to join? My ass!"

Liko closed the thumb drive, disconnected it, and set it on the coffee table.

His phone rang. "This is Liko."

"Liko, this is Danielle. We met last night."

"Hi, Danielle. How are you?" He switched the phone to speaker. "I thought your presentation was excellent." Liko said "your presentation" so Charla would know who he was talking to.

"I'm in shock," Danielle said, "like everyone else. Did you hear the news?"

"Yes."

"I still can't believe it."

"Neither can I." Liko glanced at Charla and shrugged his shoulders. *What does Danielle want from me?*

"It's terrible. Just terrible."

"Yes. It is."

There was a pause, and then Danielle asked, "Have you heard from Marimichael?"

"No."

"Neither have I." There was another pause. "I've been trying to contact her, and she hasn't answered her phone or her emails. I'm worried about her."

Liko nodded his head unconsciously. He glanced at Charla, who was sitting on the edge of the couch, totally focused on the conversation.

"I haven't seen her since last night," Liko said. "But I haven't tried to call her." He suddenly realized that he had no contact information. Only her address.

There was another pause. This time longer.

Liko ventured, "Can we get together?" He then added quickly. "I have some questions about divestiture." It was true. He had inherited a small fortune from Zahi and, undoubtedly, the monies were invested in a variety of funds, including fossil fuels. Sooner or later, when he had time, he would have to review the investments and make appropriate changes.

"I can't help you with that," Danielle said. "I'm not a financial advisor. I'm an investigative reporter who teaches journalism and occasionally a history class. I do hope, though, that you divest from fossil fuels and find an environmentally friendly fund. There are many to choose from."

Liko nodded again. "Of course." He took a deep breath, gathered up his courage, and said, "Actually, I was hoping that *you* could tell me something about Marimichael." Having ventured the request, Liko continued. "I only just met her last week."

A long, long pause followed, and then Danielle's answer. "We can meet at the Club."

"What time is best for you?" Liko asked.

"Tomorrow. Noon? For lunch?"

"Sure."

"See you then," Liko said.

When he hung up, Charla unloaded on him. "What's the matter with you?!"

"Calm down. I'm meeting Danielle for lunch, not Marimichael."

"I'm going with you."

"No."

"Why not?"

"Because you turn everything into drama."

"Liko, you need to slow down and think about what you're doing."

"Hey!" Liko stood up and raised his hands in the air, palms facing her. "Stop it!" He picked up the remote off the coffee table where she had set it. "Stop nagging me!" He clicked it, changing the screen to CNN.

A man stood in the White House Rose Garden, dressed in a suit and tie, answering questions from a gaggle of reporters. The banner at the bottom of the screen identified him as the White House Press Secretary. "Yes," he said, "it has been confirmed that the man who killed President Daniel Danson and the man who took Augustus Dimer's wife and grandchildren hostage are brothers: Shane and Fintan O'Brien."

"Damn!" Charla said, glancing at Liko.

A reporter asked, "What is known about the brothers? Is it true they both served in the military? One in the marines? The other in the army?"

"Liko," Charla said, talking over the reporter questioning the White House Press Secretary, "you need to call Jack Amaya and tell him that you went on a date with their sister. You need to give him the thumb drive."

Liko said nothing. He just backed up carefully and sat down in his Italia chair. For the next half hour, he and Charla watched breaking news, switching between CNN and MSNBC. When he finally turned off the television, he felt mentally exhausted.

8

Cosmos Club
Redux

"I taught a graduate course at West Virginia University that I called 'Dreamers and Defenders: The American Environmental Movement,'" Danielle told Liko. They were seated at a table for two in the 1878 Grille in the Cosmos Club. "We started with Thoreau and Olmsted and John Wesley Powell, and we ended with Al Gore. As you can imagine, my class was controversial."

"Why was that?"

"West Virginia is known for its coal companies—Marshall County Coal, Fairmont Coal Company, Borderland Coal Company, and the Fayette County Coal Companies. It's not known for its conservation or environmentalism."

"How about Augustus Dimer? Isn't his coal mine in West Virginia?"

"Yes, it is," she said. "Some of my students had family who worked in his mines. Like Marimichael. Some still do. But coal is a dying industry."

"Was she a good student?"

"My star student. She did very well, even though she was an undergraduate. She was fired-up about environmentalism, already an activist. She later got a bachelor of arts in political science." Danielle paused to take a bite of her salad before continuing. "Behind you, on the wall, is an original portrait of Powell," she said, pointing to a gold-framed oil painting. "He was one of the environmentalists we studied."

Liko turned in his chair and looked at the portrait. Powell's whitish-gray beard completely hid his mouth, his neck, and most of his exposed upper chest. He wore his grayish white hair brushed back neatly, revealing a broad widow's peak. He had a large pronounced nose and wide, dark eyebrows that rested above alert eyes.

"He looks like quite a character," Liko said.

"He was a remarkable man," Danielle said. "He was a soldier and an environmentalist. The Cosmos Club was born in his home in 1878."

"I think I knew that Powell was an explorer, but I didn't know he was a soldier."

"He lost an arm in the Civil War, but that didn't stop him. It didn't even slow him down. He went on to lead expeditions, and to explore the Colorado River."

"He ran the Colorado River with one arm?"

"That impressed Marimichael, too." Danielle smiled.

Liko decided that he liked the portrait, and that Powell wore no shirt or bow tie with his dark, old-fashioned coat. Liko especially liked the throbbing vein in Powell's forehead. The man had been a force to be reckoned with.

Liko turned back around in his chair. "So you knew Marimichael from your class?"

"And after she graduated. She stayed in touch, and she would stop by my office. Sometimes we had lunch together. I enjoyed our talks." Danielle paused, with a wistful smile. "She was active with environmental non-profits. She told me about different projects that she was working on. Her passion was protecting threatened and endangered species. She once told me, 'Animals can no longer protect themselves without our help. We have a moral responsibility to help them.' She felt that passionately."

Liko liked that. He felt the same way about protecting the elderly, infants and children, and disabled persons—all those who could not protect themselves. *Yeah*, he thought. *Endangered and threatened wildlife should be protected too.*

"And then one day she called and told me she was going into the army. That was a surprise. After that, we lost touch."

"When did you see her again?" Liko asked.

"A couple months ago she stopped by my office. Said she had been following my environmental blog. Like me, she was livid when Trump quit the Paris Agreement, and when he began rolling back environmental protections. Trump eviscerated the EPA. But you know that."

Liko nodded, yes, and took a bite of his poached salmon.

"Did she talk about her family?"

"Her father loved the outdoors. He was a hunter, but he also had an almost religious respect for nature. I believe that is where Marimichael got her love for nature, from her father." Danielle took a sip of her iced tea. "He worked in the coal mines. The coal dust killed him—he died from emphysema."

"What about her mother?" Liko asked.

"Her mother died an agonizing death from intestinal cancers. Marimichael believed that chemicals from coal plant wastes and ash leached into the groundwater and contaminated their drinking water. They had a well in their back yard. After her mother died, the owner of the coal mine bought their property cheap."

"Who was the owner?"

"Dimer. The coal baron."

The conversation stopped and they looked at each other across the table. Finally Liko asked the question that had been troubling him: "Is Marimichael a violent person? Could she be involved in the kidnapping? The assassination?"

"I don't know. I've never known her to do anything violent. Of course she is frustrated with the slow pace of government, with the almost total lack of engagement on climate change. But aren't we all?"

Liko nodded agreement.

"She is motivated, though. She wants to stop the sixth extinction."

"How motivated? Do you know her brothers?"

"No. Just what I read about the kidnapping. And the live footage of the assassination has been played over and over again, just like the planes flying into the Twin Towers. It is just awful!"

"Do you think she is involved?" Liko asked a second time.

"I hope not."

While they ate, Danielle told stories about one of her favorite environmentalists, Frederick Law Olmsted. She described the challenges he faced when he designed and then supervised the construction of New York's Central Park. During dessert, she told the story about how Olmsted designed the enlarged grounds of the US Capitol, too.

While she was telling her story, Liko's mind wandered. "Do you think she is one of the kidnappers?"

The expression on Danielle's face changed. She was clearly concerned—no, frightened.

She believes Marimichael is involved.

"If I hear from her," Liko said, "I'll tell her to call you."

"Thank you."

"No, thank *you* for meeting with me." He then added, "I guess we just have to wait for news."

"Well, I enjoyed our talk, Liko. We should to do it again sometime."

"I'd like that," Liko said.

"Have you seen *Soylent Green?*"

"*Soylent Green?* No, I haven't heard of it."

"*Soylent Green* is a movie from the early '70s about climate change. It describes a dystopian climate future. You should see it—it's playing at West End Theater."

"Have you seen it?" Liko asked.

"Yes, but not on the big screen. It's playing on their Classic Wednesdays, one day only."

"I'll try to make it."

9

Charla's Boyfriend

"She's worried Marimichael is involved." Liko was seated in his Italia chair, his plaid flannel robe tied tightly around his waist. He had already brushed his teeth, taken his contacts out, and gotten ready for bed. It had been a long, eventful day.

"Well, I told you so." Charla was seated on the white couch next to him with her legs folded beneath her. She had pulled the kimono as far down on her thighs as it could go. She ran her hand over one of her exposed knees in absent-minded circles. As far as Liko knew, she had spent the whole day in her kimono and had never left the condominium.

"I have the worst luck with women," Liko said.

"You fall in love too easily," Charla chided, rolling her eyes.

Liko shrugged his shoulders with a slight sigh. "Hormones? Pheromones? It's all out of my control."

Charla smiled. "Yes, that fits the pattern I see." When Liko didn't respond, she explained. "I have seen you fall in love, time after time, yet tonight you are as lonely as I am."

He looked at her in amazement. "Yes, I am lonely to the core."

"Let me tell you a story," she said. "Can I tell you a story?"

"Yes, please."

"I have been in love once. And by love, I mean head-over-high-heels passionately in love. I was fifteen and he was two years older, seventeen. I was a sophomore and he was a senior." Charla paused and took a sip of her wine, then readjusted her kimono.

"He loved me too. Yet... not me... the real me...." Charla looked to see if Liko understood what she was saying. She repeated herself, just to make sure: "He didn't love ME, because he didn't know ME."

Liko nodded his understanding.

"We were quite the topic in our high school. It didn't help, of course, that I shaved off my eyebrows my junior year. Isaac—that was his name, Isaac—he didn't like it either. But I had bought eye makeup and I wanted to try it." Charla paused and looked at the window for a long moment, lost in the memory.

"But back to my story." She bit the lower corner of her lip. "I'll try to keep it short. I've noticed that ever since you met this girl, Marimichael, your attention span has grown shorter and shorter. Hormones, Liko, hormones. I have no doubt there is a chemistry between the two of you. You need to recognize, though, what you are 'feeling' is 'chemistry.' Yes, it's real. But it is 'chemistry.'" Charla turned her body towards Liko and reached out to touch his arm.

"My friend, you have a choice: Who is in control? You, or the chemistry? Isaac loved me. Yet, like I said, he didn't know *me*. We were good in bed. Or at least I pleased him and that made me very happy. And he thought I was handsome. Of course HE was the most handsome boy in our school." Charla smiled.

"After we graduated, I began my sex change." She adjusted herself on the couch. "This is who I am, Liko." Charla held her hands up towards the ceiling and spread them wide.

Liko looked at her and again nodded his understanding. *Where is she going with this?* he wondered.

"Isaac didn't understand. He watched me slowly change — physically — from the body of a man to a woman. He didn't understand that *this* is the true me, the real me. I'm a woman, not a man. I never have been. But Isaac didn't understand. It broke his heart. And mine, too." Charla looked down and clasped her hands together, lost in thought. Liko thought she was so beautiful.

"Isaac had fallen in love with someone who did not really exist. An illusion. He saw what he wanted to see. He saw my skin and hair and lips, but he never saw what was in my eyes. What was in the core of my being. I'm Charla." She abruptly turned toward Liko and looked him directly in the eyes.

"I'm afraid for you, Liko. Your vision of Marimichael is not Marimichael. It is not who she is—"

Liko held up his hand to cut Charla off. They gazed into each other's eyes.

"I love you, Liko," she said. "You are becoming my best friend. I don't want to see this woman hurt you."

"I imagine that you will scratch out her eyes if she does?"

"Yes, I will!"

The intercom buzzed. Liko jumped up from his chair as if he was quickly rebounding from a punch. He stepped to the security panel and gently pushed random buttons as he had before. "This is Liko."

"A man from the FBI is here to see you."

"At this hour?" Liko said to Charla. They exchanged glances.

"Please send him up," Liko told the concierge at the front desk.

Jack was wearing a light purple shirt and a darker purple tie. It was tasteful. He had on black pants and a dark, navy blue sport coat.

"Come in," Liko said. "Have a seat."

Jack offered his hand and Liko shook it. Jack then stepped between the kitchen island and dining table and into the living room. He sat down in Liko's Italia chair.

Liko glanced at Charla and saw her response: she smiled as she once again took the measure of "Jack the FBI man."

"I am here to ask for your help. Shane has demanded that you be his intermediary. I hope you will reconsider."

"I'm still not interested," Liko said.

"Shane believes that we, the FBI, are stopping you. He believes that we, the FBI, won't allow you to be the intermediary." Jack paused for his words to sink in. "He thinks we are responsible, that the FBI is not meeting his demands. Not you."

"I'm sorry," Liko said, "but I'm still not interested. Like I said: this is outside my pay grade."

"He has threatened to destroy a bridge into DC." Again Jack paused. His eyes did not move from Liko's face.

"I don't understand," Liko said.

"He says he will destroy a bridge into DC." Jack looked down at his feet and then at Charla and then back to Liko. "We have reason to believe that he can do it, that he has the means."

"Shit!" Liko stood up and paced back and forth between Jack and Charla. He felt trapped. The room seemed smaller. It was becoming claustrophobic.

He took a quick breath. "There is no other way?"

"No. Shane O'Brien is very specific in his demands."

"He can blow up a bridge?" Charla asked.

"We think so."

"Damn!" she said.

"Is his brother Fintan the man who assassinated the President?" Liko knew the answer because he had heard it on CNN, but he still wanted confirmation.

"Yes, he is."

"Are there others?"

Jack considered the question for a moment before answering. "Yes. And they are well organized."

Liko looked at Charla. She was gesturing at him. *I know she wants me to tell Jack what I know about Marimichael, but I'm not going to do it.*

Jack noticed Charla gesturing with her hands. "Did I miss something?"

"No," Liko said, quickly.

Charla made a disgruntled face at Liko.

Liko suddenly felt angry. *Why? Why am I so angry?* He sat down and took several long, deep breaths. He focused on the newspaper on the coffee table that was still open to the headline: "President Danson Assassinated." He took another deep breath. *I'm scared. For the first time in a long time, I'm scared.*

He shifted his attention back to Jack. "What would I have to do?"

"We'd have your back," Jack said. "The FBI. All of us. Everyone."

"That's not what I asked."

"We'll prep you before we send you in."

"OK, I'll help, but with reservations. What do I need to do?"

10

Personal Message

Liko recognized the mournful ballad playing over the music system in Dimer's mansion: Neil Young's "Green is Blue" from his album *Colorado*.

"Why did you ask for me?" Liko asked Shane. "Why are you doing this to me?"

"You're famous because of the Davos affair, yes? People are interested in you. They'll listen to you."

"Why not just email or text message your demands? Or just tell the press?"

"Because I have something for you to deliver in person to the oil baron."

Liko's eyebrow raised. "What?"

"Locks of their hair... the hair of his wife and his grandchildren."

Liko saw a look of satisfaction in Shane's face.

"There are two messages for Augustus Dimer: one public, one private."

"The public message?"

"Shut down your coal plant. The old one in West Virginia – Dimer Coal."

"OK. The private message?"

"When you hand Dimer the locks of hair, whisper a message in his ear."

Liko couldn't imagine what Shane might say. A private message? "What do I whisper?"

"Stop the filibustering. Tell the Senate Minority Leader, Senator Nappe, to stop filibustering. Allow a vote in the House."

Liko's eyes widened. "You think the oil baron owns the Senate Minority Leader?"

"Of course he does," Shane stated flatly. "His dark money owns the Republicans in the Senate." Shane handed Liko a file.

"What's this?"

"A fact sheet about Dimer's coal plant: toxic chemical releases into the air, land, and groundwater; CO_2 emissions; disease and death trends of employees."

"For the press?"

"Yes. You will find a press release inside the file, too."

"You prepared all this ahead of time? Before the kidnapping?"

Shane smiled at Liko's naiveté. "Of course. I hope you will read the press release publicly. Will you do this for us?"

"Who is us? Does that include your sister?"

Shane looked at Liko, smiled again, but did not answer him.

"Why me?"

"Because you have credibility. The press will recognize you and identify you as the man who stood up and spoke the truth at Davos. The public will listen to you."

"So you want me to give Augustus the locks of hair, whisper your message in his ear, and read this press release. Anything else?"

"No."

Liko considered the request for a moment. "Why should I?"

"Why? Really?"

"Yes. Why should I help you?"

"Because you want the Carbon Pricing Bill to pass. Because you want to see CO_2 emissions decline. Because you want to save humanity. Because you do not want to see mass extinction during your lifetime."

"I'll need to read it first. Make sure I agree with it. I won't deliver it unless I agree with it."

"Liko, it's your choice whether to read it publicly. But don't fail to deliver the private message to Dimer. He is holding Congress hostage, yes? Well, I am holding his family hostage. Either he allows the climate crisis bill to come to a vote on the floor of the Senate, or I will kill his family."

Liko had no doubt that Shane would kill them.

"But why the private message? Why not just say it publicly, too?"

Shane looked intently at Liko. "You can buy a senator, but the public doesn't like you to blackmail one."

"So why don't you work within the system and buy Senator Nappe," Liko asked, "like the corporations and Augustus Dimer do?"

Shane slapped Liko.

Liko wanted to strike back but fought the impulse. He

felt his legs trembling with anger. "What the hell was that for?"

"You are being naive."

Liko stared at Shane as he struggled to control his anger.

"One last thing," Shane said. "You asked about Marimichael?"

"Yes."

"She thinks you are awesome. What you did at Davos. Punching out the kleptocrat Derichenko. She watched that video again and again. I was impressed too. Liko, you now have the opportunity to do something even greater: help us jump start the climate crisis revolution."

The side of Liko's face burned where the blood vessels had been broken.

The two men stood facing each other, neither saying a word, neither moving. Finally, Shane said, "I apologize for striking you."

"Apology accepted."

"Liko, let me give you some advice, something I was taught in the marines: Visualize how you are going to handle a situation before it happens. Decide to be confident and visualize success."

That is exactly what I'm doing for the FBI. Studying you and the other two 'eco-terrorists.' Here I am, right where I don't want to be, right in the middle of everything. Out of my comfort zone.

"Is Marimichael involved?" Liko asked again.

"Join us?"

"Your brother assassinated the president, and you want me to join? President Danson was a good man. His record was pro-environment. Why assassinate him? That makes no sense."

"He believed in miracles, Liko, and we don't."

"Miracles?"

"He believed in carbon capture and sequestration. Miracles."

"So you assassinated him?"

"He planned to invest in geoengineering, heavily. Geoengineering is a last resort of the most dangerous kind."

Liko wasn't sure what Shane meant by 'geoengineering' or 'last resort of the most dangerous kind.' It did sound more science fiction than science, though.

"We need wind and solar," Shane said, "not unproven technologies."

Liko was confused but said nothing.

"Join us," Shane said. "Our revolution will succeed."

You have got to be kidding, Liko thought. "You know they're going to kill you. You won't leave this mansion alive." He couldn't help but say that, but he immediately regretted it.

"I knew that when I knocked on the front door."

"You knocked?"

"Yes," Shane said.

"Actually, we rang the doorbell," the man standing behind Liko said. He had an automatic rifle aimed at Liko. *A Green Beret like Shane?*

Liko had observed a third rebel moving about the mansion, now and then passing by the door to the library where Shane and he were meeting. The three men appeared to be a disciplined, tight group. *They have a history together*, Liko thought.

Liko turned around and said to the burly man holding the automatic rifle, "They will kill your ass, too."

Liko couldn't believe he said that even as the words came out of his mouth, but he was pissed at being thrown into the current situation, caught between the FBI and a group of terrorists, especially since he believed in what Shane and the rebels hoped to achieve, just not how they were doing it.

"One last thing," Liko said. "The FBI wanted me to see the hostages. Make sure they were OK."

11

Locks of Hair - Heidi

"Fools. All of you are fools."

Liko had no doubt that she meant it. He could hear the conviction in her voice and see it on her angry face.

"Augustus would just as soon see me dead — and lose his grandchildren — than close one of his coal plants." She gestured towards the man holding the automatic rifle. "You might as well shoot us now." She sat down in a chair. "And none of you will leave here alive either. Fools. All of you are fools!"

Liko was taken aback at the vehemence of her statement. He wondered if she was being sarcastic or dead serious. She looked like she was unharmed, though.

"Do you need anything?" Liko asked.

"Pssst!" she said, guffawing. "Who are you?"

"Liko Koholua. The FBI sent me to see that you were OK. Are you OK?"

"Physically, I'm fine. The grandkids, too." She looked closely at Liko. "I do hope that you can get them out alive."

Liko looked at the angry woman in front of him. *I'll do everything I can.* Turning slightly, he said, "Shane, would you mind leaving us alone for a few minutes?"

"Sure," Shane answered. "I don't see any harm in it."

He stepped out of the recreational room and locked the door behind him. Liko suspected that at least one of the eco-terrorists remained just outside the door. There were no windows or man-size vents in the room, so no possibility of escape.

"I'm just a civilian stuck in the middle," Liko told her. "I'm not with the FBI. And I'm definitely not working with these men."

"My name is Heidi," she said.

"Sorry to meet you under such bad circumstances."

"That's all right," she said.

"Is there anything you want to tell me?"

She thought for a moment and then said, "He's an asshole."

"Who is an asshole? Which one of them?"

"No. Heaven's no. Not these men. I'm talking about Augustus, my husband. He's a certified narcissistic, cowardly asshole."

"I see." Liko thought her denouncement of her husband was astonishing, especially under the current circumstances, but decided to remain silent.

"He only married me because I was a model and beautiful. I was too naïve to realize his motives at the time, but that's the way it was. He never loved me."

"I'm sorry to hear that."

"But that's not the worst of it. He doesn't respect me. How can you live with someone day-in and day-out if they don't respect you?"

Liko felt embarrassed at the intimacy of her statements. "I don't know," he said.

"Sadly, it took me a long time to realize that THAT was the real trouble in our marriage: lack of respect." She pushed her hands slowly down her slacks, from her thighs to her knees.

"You see," she said, "I'm less educated than Augustus. I only have a high school diploma. He has advanced degrees in chemistry and petroleum engineering. He thinks I'm stupid. But I'm not."

"Perhaps if I had given him a boy. But as fate would have it, I gave him a girl, my Candy."

Liko smiled at the name of her daughter.

"That's just her nickname, Candy."

"I see."

"I never worked outside the home. Never had a job or career. Just raised Candy. She's somewhere in Europe now. Maybe she heard about our kidnapping. Maybe she's on her way back home. Do you know?"

"No," Liko said. "I'm sorry. I don't have any information about your daughter."

"Her two kids, my two grandchildren, are here with me watching television or playing video games in the next room. They're fine. No one has touched them." She reached up absentmindedly to touch her necklace.

"Let me tell you about Augustus. He teases people. He enjoys it. He doesn't stop until he has you crying."

Liko could see her eyes beginning to water. He surmised she was reliving painful memories.

"When we were first married, he pushed me out of bed

onto the floor with his feet. One time he poured a Coke on me and laughed." Heidi's face shifted from pain to anger.

"Now that these men have come into my home, Augustus has an excuse to kill me. He loves money and his coal plants. Not me, not his grandkids, not anyone. Only himself. Like I said, he's a narcissist. He only loves himself."

The conversation went on like this for another five minutes. She had nothing good to say about her husband, but despite everything she was saying, Liko suspected she still loved him. She hated him and loved him. *Relationships can be so confusing.*

Finally Liko said, "I'll tell the FBI what you said."

"Nothing they can do," she said. "He loves his money. Not us."

Her resignation to her belief that she would soon be killed was unsettling.

"Can you protect my grandchildren?" she asked. "Please. I beg you to protect them. Promise me you'll do everything you can."

12

Debriefing and Soylent Green

Liko gave Jack the folder that contained the fact sheet and press release about the Dimer Coal Plant and the envelope with the locks of hair. He explained Shane's demands to Jack. "Dimer has to meet me at his coal plant. I have to hand him an envelope and read a press release, and then I'm supposed to whisper a personal message from Shane."

"What message?"

"Dimer must tell the Republican Senate Leader to stop filibustering the climate action bills. He must allow a vote on the bills."

Jack slapped the rolled-up folder against the side of his leg as he crossed the floor to another FBI agent. The man was dressed like a typical FBI agent, unlike Jack. He looked like a Mormon in his black suit, white shirt, and dull tie. "Check out these materials. If you find anything of concern, report back to me." He passed the envelope to the agent. "Right away."

"There is more," Liko said, summoning Jack.

"What?"

"Dimer's grandkids are OK, but his wife, Heidi, is afraid of him."

"How so?"

"She thinks that he will take advantage of the kidnapping to kill her." Liko then told Jack everything Heidi had said.

Jack did not seem surprised. "I knew he was a narcissist. Everyone does. But I didn't know that he abused his wife psychologically."

Liko thought about that. Heidi Dimer lacked self-esteem, something Augustus had denied her through the years. Consequently, she was rich but terribly unhappy. She also doted on her grandchildren, most likely because she was lonely and craved affection.

"One more thing," Liko said. "Shane said the president was assassinated because he was deceiving the public. It had something to do with carbon capture and sequestration. Shane said the president had been in the pockets of the fossil fuel industry. He was deceiving the public and working with the fossil fuel companies."

"I see," Jack said.

"Do I have to stay here?" Liko asked.

"No," Jack answered.

"Then I would like to go home."

"OK. We'll pick you up first thing in the morning, at five o'clock."

"I'll be ready."

As soon as the FBI driver dropped him off at his condo,

Liko called Danielle. "Are you still going to see *Soylent Green*?"

"Yes," Danielle answered.

"May I join you?"

Pause. "That works. I'm close by, at the Cosmos Club. I need dinner, though."

Liko looked at his watch. It was six o'clock and already dark outside. The last show didn't start until 9pm, so they had plenty of time.

"How about Rasika?" he suggested. "It's just around the corner from my condo and a block or two from the theater."

"Sure," Danielle said, "I like Indian food."

She added, "While we dine, I'll give you a summary version of my crash course in environmental activism. I spend a whole week on activism in my environmental history class."

"That sounds great. You can park in my building in a guest parking place. I'll arrange it with the front desk. It's an easy walk to Rasika and the West End Theater."

"Sure," she said.

Liko wrote a note for Charla: *Gone to dinner with Danielle. Be back late.* He set the note on the corner of the kitchen island so it would catch her eye.

Liko was amazed that there was such a great restaurant on the same block as his condominium building. He made reservations through the condominium's concierge, who told him Obama had favored the restaurant when he was president.

Danielle arrived with a book for Liko: Edward Abbey's *Desert Solitaire*. She had checked the book out from the Cosmos Club library.

Turning *Desert Solitaire* over in his hands, Liko asked, "Have you read Marimichael's book about John Brown?"

"No."

Liko tried to recall the title. Marimichael's book was on the thumb drive that she had given him, in pdf. "I believe the title is 'The Legacy of John Brown and Climate Change Revolution' or something like that."

"No. I haven't read it." Danielle's voice betrayed her surprise. "She didn't share it with me."

"Well," Liko said. "I haven't read it either." He wondered when he would have time. And now he had to read Edward Abbey's book, too. "I have an electronic copy I can send you."

"Like John Brown, the O'Briens have made a tragic mistake." Danielle sipped her sparkling wine. "Kidnapping creates negative feelings among the public. It's a bad choice for revolutionaries to make."

"I don't think Shane will hurt Heidi Dimer or her grandkids."

"John Brown didn't hurt his hostages, either. Nevertheless, he placed them in danger, and that is totally unacceptable to Americans. I personally think it is reprehensible. I fear it will not end well."

"So do I," Liko said.

They ordered the four-course tasting menu. For their third course, Liko chose the black cod and Danielle selected the lamb pepper masala. While they ate, Danielle shared her summary of environmental activism in the United States. Liko found it fascinating.

When it was time for dessert, Liko changed the topic to current politics. "Do you know anything about the Republican Senate Leader?"

"Senator Nappe?"

"Yes."

A waitress brought out two date and toffee puddings. They each took a spoonful and tasted it, slowly.

"I've interviewed him. He has no real ideology. The fossil fuel industry has always bankrolled his campaign. They keep him in office. His core values are anti-government, anti-regulation, and anti-environmentalism. He's a racist and a white supremacist."

"This is delicious." She paused to point her spoon at the pudding. "He revels in the power of his position. He has a huge ego. He thinks that he is wise and that Americans are stupid and deserve to be manipulated."

She spooned up another bite, savored it, and then pointed the silver spoon at Liko. "He is heartless, vain, and in need of constant adoration."

"Anything else?" Liko asked. He pushed the small plate with the remaining toffee pudding towards Danielle.

"Yes." She swirled the last of the pudding onto her spoon and took the last bite. "He pretends to be a man of god, but he is a charlatan, theatrically pious, and a hypocrite."

"I take it that you don't like him?" Liko said, sarcastically.

"He is terrible!"

"Is there any connection between him and Augustus Dimer?"

"No doubt. Most Republicans are puppets of the fossil fuel industry."

They finished eating and Liko paid for the meal. He left a large tip.

As they walked to the entrance, he saw Charla sitting at the bar in the front of the restaurant chatting with two men, one on each side of her. She was wearing a white poplin shirt with a high neck and puffy shoulders tucked into a cheerful orange skirt that was ruffled at the bottom. Charla was charismatic and attractive.

"Hi," Liko said, pausing with a smile.

"Hi."

"This is Charla," Liko told Danielle. "A friend of mine." Danielle nodded graciously to Charla and the two men.

"From the Cosmos Club lecture?" Charla asked Danielle.

"Yes," Danielle said.

"I'm sorry I missed your presentation. I heard it was good."

"Thank you," Danielle said. "I'm giving the lecture again next week at the Smithsonian Ripley Center."

"I'll try and make it," Charla said.

"I hope to see you there," Danielle said.

Liko winced, slightly. "We're on our way to the movies to see *Soylent Green*."

"Enjoy yourselves," Charla said, returning her attention to the two men.

Liko and Danielle left Rasika, crossed M street, and walked two blocks to the theater.

Soylent Green enthralled Liko. He identified with the main character Thorn, an honest detective played by Charlton Heston. Liko identified with Thorn's determination, his physical strength, and his recklessness. He imagined himself as Thorn, investigating the assassination of a wealthy man who had profited from the production of Soylent Green, a mysterious food bar. With the help of Saul, a researcher played by Edward G. Robinson, Thorn uncovered the dark truth about the green blocks of food: it was made from the bodies of dead people. The revelation shocked Liko.

Near the end of the film, Liko watched as Thorn's friend Saul swallowed a lethal potion and lay down on a comfortable bed to die in a euthanasia center. As Saul listened to Rachmaninoff, a wrap-around movie showed the natural world as it used to exist before an ecological disaster destroyed Mother Earth. Liko watched Thorn, as Thorn watched his friend Saul die. Liko thought Edward G. Robinson's death scene was worthy of an Oscar. Later, Danielle told him it was the actor's last performance.

The ocean waves and waterfalls in the wrap-around movie reminded Liko of the beauty of Hawaii. He remembered schools of colorful reef fish and a manta ray that had rubbed against his bare skin. He remembered sitting on a beach in Waikiki and watching the sun set into the Pacific Ocean. He remembered sighting a 'I'iwi, a bright red honeycreeper with an orange-red bill, in a remote Hawaiian forest.

But then Liko recalled that the number of surviving 'I'iwi had plummeted recently because warmer air temperatures had increased the habitat of the mosquitoes that fed on the birds, spreading avian malaria and avian pox. Honeycreepers and reef fish now faced extinction because of global warming. The oceans were warming, becoming more acidic, and the reefs were dying worldwide. Like the Great Barrier Reef in Australia, the coral and animal life in Hanauma Bay were disappearing.

Later Danielle would tell Liko that Soylent Green had been filmed more than 50 years earlier. It had predicted the current destruction of nature, worldwide, because of climate change. Scientists had warned the oil companies, the fossil fuel industry, and the politicians more than fifty years ago and they had done nothing.

Liko leaned back in the theater seat. He felt depressed and angry. *They've been playing us for fools. All these years!*

After escorting Danielle back to her car, Liko returned to his condominium to find Charla waiting up for him. She had changed into her kimono. It was approaching midnight.

"I couldn't sleep," she said.

"I doubt that I can sleep either," he said. "I'm going to fix a mug of hot lemon water and honey. Would you like anything?"

"No thanks," she said.

Liko fixed the sour drink and sat down in his Italia chair adjacent to Charla on the couch.

"What is she like?" Charla asked.

"Danielle?"

"Yeah."

"Smart. She used to work for *The Baltimore Sun*. Now she's a freelance journalist. She has her own blog."

"She lost her job at the paper?"

"She said she followed the editor-in-chief out the door when he was fired. Now she teaches classes in journalism and history to pay her bills."

"She likes what she does?"

"I think so. She gets to ask questions and figure things out."

"And then share it with people?"

"Yes. Not a bad job, yeah?"

Charla nodded.

Saul's death scene at the end of *Soylent Green* kept looping endlessly in Liko's mind. Like Saul, he couldn't believe what mankind had done to themselves. But unlike Saul, he had no way to escape.

"Shane wants me to give Dimer a private message."

"No kidding? What is it?"

"You must not tell anyone. You promise?"

"Of course." She nodded her head, agreeing to keep the secret.

"I'm supposed to whisper in his ear: 'Stop the filibustering. Allow a vote on the climate crisis bill.'"

"What?!" There was a moment of silence and then Charla said, "You told the FBI?"

"Yes."

"It's blackmail, Liko. Pure and simple. Shane is blackmailing a Senator, do this or else."

Liko sank down in his chair in frustrated agreement. "But the Senator won't allow a vote on the bill."

"I understand your concern, your frustration, your anger, but he has the right to filibuster. The filibuster is an instrument — a tool — of our democracy."

"Bullshit!" Liko said, raising his voice. *The oceans are dying, doesn't she get that?* "The fossil fuel industry has bought Congress, don't you understand? What Dimer has done is worse than blackmail. He holds not only Senator Nappe but our system of democracy hostage, and the price that's being paid is the complete destruction of the natural world! Don't you understand?"

Now it was Charla's turn to say, "Bullshit! What Augustus has done is not illegal, even if I disagree. He is not holding a grandmother and her grandchildren at gunpoint. Is he?"

"Worse," Liko said. "He is holding the whole country hostage. He has paralyzed our government, even though a majority — a large majority of the public — want change. They want immediate action on the climate crisis, but he won't allow a vote."

They argued, raised their voices, and pleaded contrary positions.

Charla said, "Did you learn anything about Marimichael? Is she involved?"

"I'm tired of your badgering about her."

"Then pull yourself together!" Charla raised her voice. "Grow up! Don't be stupid."

"What?"

"You heard me," she said. "Sometimes you lack good judgment."

"That's it." He was exhausted and her words cut deep. "I've had enough."

"Did you give Jack the thumb drive?"

He hadn't, and he knew that she knew it because it was on the coffee table between them.

"There's nothing useful on it." He stood up and stared at Charla. "It's just a recruitment tool."

She stood up and stared back at him across the coffee table. "Really? How do you know that?"

Earlier he had decided to give the thumb drive to Jack, but with all the stress he had forgotten. And now he felt concerned about it. Would it get him into trouble?

"I wish you would stop nagging me," he said.

"Grow up, Liko!"

"You are so damn annoying," he said.

He saw the impact of his words on her face. He saw the slight quiver in her body and then she burst into tears. Her body began to tremble. She stood in front of him as if riveted to the floor, tears streaming down her cheeks.

Liko saw the pain he had inflicted, nevertheless he yelled at her: "Leave me alone! Stop interfering!" He turned and walked into his bedroom and closed the door hard.

He sat down on his bed. "I don't even like you!" he said, raising his voice so she could hear him.

A half minute passed and then a minute. In was late, after midnight. He sat on the end of his bed, sulking, then he stood and undressed. He brushed his teeth, avoiding his reflection in the bathroom mirror above the two hand sinks and along the full length of the countertop. He splashed warm water on his face and dried off with a soft white hand towel. He returned to the bed and slipped naked between the soft sheets.

His mind raced. *How can I protect Heidi? And her grandchildren?* He tossed and turned.

Suddenly he got up and slipped into his flannel robe. He went into the living room to apologize, but Charla wasn't there. He went to her bedroom. The door stood open. He rested a hand against the doorjamb. It was dark and he could not see inside.

"Charla," he said softly. "I'm sorry. I was a real asshole. I don't know what came over me. Please forgive me. I love you. You know that I think you are wonderful."

There was no answer. As his eyes adjusted to the dark, he saw her bed was empty.

Damn! he thought. *She's gone!*

And then he saw her on the porch just outside her room. The porch door was closed to keep the air-conditioning from escaping.

He started to cross her bedroom to the porch but thought better of it. What would she think of him entering her bedroom without first asking? *It might make things worse.*

He started to call her name to get her attention, but he hesitated. *I'll apologize in the morning*, he told himself.

13

Helicopter Ride

Mid-morning the next day, Jack and Liko climbed aboard a large military helicopter purchased by the FBI. *Overkill,* Liko thought. *Excessive force.*

The passengers included members of the FBI Crisis Negotiation Unit out of the FBI Academy in Quantico, Virginia. The men looked like commandos. One carried a rifle and Liko guessed he was a sniper. Another had an arm patch that read "Operational Medicine." In the center of the patch was a snake wrapped around the blade of a sword.

One FBI man sat apart from the team. He wore conservative apparel: a dark coat, dark suit, white shirt, and striped tie. An American flag pin waved from his lapel. He had tucked his narrow-striped tie between the buttons on his white shirt so the tie did not flop around with the air currents. Jack wore a bow tie. It looked like it was made out of feathers. *Very eccentric, as always,* Liko thought.

"You've got a good partner, here," the FBI man told Liko, tipping his head towards Jack. "No one can wrestle numbers and pin down suspects like Jack." He smiled at Jack.

Jack shook his head at the compliment.

"He's one hell of a forensic accountant," the FBI man added. "He's taken down several corporate executives and their companies for tax evasion."

"It's always a team effort," Jack said. "I direct the Bureau's Public Corruption program in DC."

"He's been after this son of a bitch Dimer for years. And what happens? Some asshole eco-terrorist kidnaps the man's wife and grandkids. Who'd have imagined?"

Liko glanced at Jack. So that was his connection to the case: he had already been investigating Dimer before the kidnapping.

"Ever ridden in a Black Hawk?" the FBI man asked Liko.

Black Hawk? Liko thought. *Damn!* "No sir," he said.

"You're part of the professor's team, now," the FBI man said. "That's what Jack's friends call him back at the office: 'The Professor.'"

"Why?" Liko turned to see how Jack felt about the nickname.

Jack had looked away and was perusing his cell phone.

"Because," the FBI man said, "if you get Jack started, he'll talk your ear off about the US Constitution. Like he was teaching a class in American government or something." The FBI man smiled at Jack, who was focusing on his cell phone. "My name's Franz," the FBI man said. "And you?"

"Liko."

"Pleased to meet you, Liko."

Franz held his arms out to his sides and then tipped them gently up and down as if soaring. "I love flying," he quipped. "How about you, Jack?"

Jack looked up from his cell phone. "I like my feet on the ground."

"He dances, too," Franz said. "A man of many talents. Huh-huh-huh."

Liko knew that Franz was teasing Jack, but it was also clear that he respected him. Franz's laugh, though, was annoying.

"Flamenco, I think," Liko volunteered.

"I've seen him merengue," Franz said, grinning. "Through the lobby of a Four Seasons carrying a bucket of ice and a bottle of Champagne. Huh-huh."

"You'd best stop," Jack said, "before I start telling stories about YOU."

"Remember, whatever you do, don't get Jack talking about the US Constitution!"

Liko nodded.

Franz added. "I have no doubt that he has memorized it, word for word."

"I have," Jack admitted. "I know it by heart."

"And the Declaration of Independence, too?"

"Yes." Jack looked up from his cell phone.

"I imagine you've memorized the United States Tax code, too," Franz teased.

"Parts of it."

"Huh-huh-huh."

Liko realized why he disliked Franz's laugh: it sounded condescending. And the way he looked at Liko when he laughed, as if a mutual understanding had been shared, yet unspoken—yeah, it was condescending.

Liko looked closer at Franz. His complexion was fair and his eyes blue. The ends of his short blond hair wanted to curl. If his hair were a little longer he'd look like a Greek god. But his laugh betrayed him. A ladies' man? Undoubtedly. An athlete? Probably. A man who was faithful and could be trusted? Maybe, maybe not.

The helicopter banked sharply to the right, creating a temporary lull in the conversation as the men's attention was diverted to quick glances out the windows.

Liko wondered if there were openings at the FBI academy, but then he smiled to himself. *I'd never pass the background check! Huh-huh-huh!*

The Dimer Coal Plant came into view. Black smoke rose from several tall chimneys, forming a huge gray plume of condensed water vapor and particulate matter. The passengers watched the plant spew tons of CO_2 into the air, right before their eyes. They were all witnesses. The helicopter circled the butt-ugly concrete towers.

The hyperboloid shape of the cooling towers astounded Liko. The towers were the same shape as Hawaiian ipu heke gourds. Liko had seen his great-aunt's collection on display in her home in Hawaii. Craftsmen had created the percussion instruments from two gourds of different sizes, joining them together at their necks.

Liko recalled his Uncle Keahi dancing and chanting to the beat of an ipu. Liko tenderly remembered a cookout that his Uncle Keahi had thrown for friends around the pool of his condominium in Oahu. Uncle Keahi had held the ipu heke by the neck, between his legs, as he sat and chanted.

Liko watched the huge gray plume rise from the cooling towers and empty into the air, as if the sky were an open sewer. He knew that the Dimer Coal Plant had polluted the groundwater beneath the plant with heavy metals of selenium, cadmium, mercury, and arsenic. Heavy metals had leached from the coal waste and coal ash that were stored onsite. That information was from an EPA study cited in the fact sheet Shane had given Liko to distribute to the press. Men, women, and children had died from groundwater contaminated with arsenic.

The sight of the coal plant cemented Liko's determination to convey Shane's private message. He recalled the tense discussion he'd had the night before with Charla about the Senate filibuster. She had been asleep this morning when the FBI driver picked Liko up in front of the condo before sunrise. *I wish I'd had time to apologize.*

Now, with a bird's-eye view of the coal plant stacks, Liko recalled a scene from the movie *Doubt*, when a priest tells a gossipy woman to cut open a pillow on the roof of her house and shake out the feathers. The woman does what the priest asks, and feathers fly everywhere. She then returns and asks the priest, "What now must I do?" The priest tells her to return to the roof and collect all the feathers and put them back into the pillow. Of course, the woman is astonished because she knows that the request is impossible.

Looking at the coal plant stacks, Liko wondered how mankind could collect all the CO_2 emissions and return them to the earth. *Impossible,* he concluded. Seeing how Dimer used the atmosphere like an open sewer filled Liko with despair. *Instead of whispering in his ear,* Liko thought, *maybe I'll bite off a piece of it, just like Mike Tyson took a piece of Evander Holyfield's ear.*

The helicopter descended into a field a quarter-mile

distant from the coal plant. The back end dipped and then it leveled out, and the FBI tactical aviation pilot made a smooth landing.

A few minutes later another helicopter approached. "A McDonnell Clincher Little Bird," Jack said, slapping his thigh.

"Augustus Dimer?"

"I think so. It's a private helicopter."

14

Coal Speech

There was nothing remarkable about the appearance of Augustus, or his son, Gus. They looked like everyone else, and that surprised Liko. Weren't they supposed to be god-like, or at least larger than life? After all, they were billionaires. But they looked like any other person Liko had ever met.

There was nothing physically distinctive about them. The son, though, had similar features to the father: grayish-blue eyes, a receding hairline, and thin lips. Both were average in height.

As they were introduced, Liko stepped forward. He towered over them. He handed Augustus the small, sealed envelope that contained his wife and grandchildren's locks of hair.

Dimer tore open the end of the envelope and glanced inside, then folded the torn end of the envelope and slipped it into his inside sport coat pocket.

Liko stepped forward, leaned in, and whispered Shane's message into Augustus' ear: "Stop Senator Nappe's filibustering. Allow a vote on the climate crisis bill."

As Liko stepped back, Gus stepped forward and stiff-armed him, causing him to stumble sideways several feet. Liko regained his balance and turned to face the son.

"Never approach my father without his permission," Gus said with a scowl.

Startled, Liko stared at Gus. Liko knew that he could knock him out with little effort. He smiled remembering a similar fantasy about President Trump—being in line to shake his hand during some imagined ceremony and contemplating whether he should punch Ol' Bone Spurs or not. Now, faced with a similar real-life opportunity, Liko decided not to hit the coward.

Liko's smile, however, irritated Gus, who responded in an angry voice: "Don't mess with us, asshole."

Liko took a step away from Gus, hoping the gesture would pacify the banty rooster. He had no reason to argue or fight. He then stepped up to the gaggle of reporters in front of the coal plant. He read aloud Shane's public message about the plant, staying on script. He had perused the message earlier and agreed with it, word for word.

He then handed out copies of Shane's fact sheet to the reporters. Liko knew that the FBI had reviewed it and had allowed its distribution.

Members of the press showered Liko with a barrage of questions, and he politely deferred to the FBI public information officer who stepped forward. Several reporters began asking Augustus questions, too.

Augustus punctuated his answers with frowns, angry eye contact, and low-volume cussing. He became frustrated with their questions, and in anger, he tore off their microphones from his sports coat and threw them to the ground. "Fake news!"

The reporters moved in closer and surrounded Augustus so the directional mics on their cameras could pick up his responses.

Liko saw Gus say something to a member of his security detail, who issued orders and bodyguards ran forward, shoving and pushing the reporters aside. Bodyguards spat into camera lenses, grabbed expensive cameras, and smashed them on the ground.

They grabbed a sign from an elderly woman protestor and ripped the poster board. Liko lunged in her direction to help her but was cut off by a member of the FBI swat team—the man with the arm patch that read "Operational Medicine." Restrained, Liko watched as the woman stumbled and fell to the ground. A bodyguard kicked her repeatedly as she lay in a fetal position until a member of the SWAT team forced his way between them.

Liko pulled out his iPhone and began taking pictures. By the time he started recording, the FBI swat team had separated the bodyguards from the reporters and peaceful protestors.

"Dammit," Liko said. "I wasn't fast enough."

The FBI SWAT team member relaxed his grip on Liko. "Don't worry, sir," he told Liko. "I'm sure a reporter captured all on video. It's stupid to attack reporters while they're shooting."

15

Pumpkin Bread

When Jack dropped off Liko in front of his building it was late, around eleven o'clock at night, yet the Ritz Carlton across the street was still busy. Black SUVs pulled up, one after another, to pick up fares. *An event must be letting out,* Liko thought.

"Welcome back Mr. Koholua," the doorman said, holding the door open for him.

"It's good to be back," Liko responded. He walked across the lobby and took the elevator to the sixth floor.

He found the door to his condominium unlocked. *It's probably just Charla forgetting to lock the door again,* he thought, *which means she's here and I can finally apologize.* He opened it quietly, flicked on the ceiling lights over the kitchen island, and entered. He grabbed an apple off the bowl of fruit and took a large bite. His stomach growled.

He found Charla asleep on the couch, her head resting on a cushion, her legs outstretched, and a glass of red wine tipped in her hand. He set down the apple on the coffee table and gently pulled the wine glass from her fingers. He then walked, quietly, back into the kitchen and set the crystal glass next to the sink, carefully, making no sound. He flicked off the recessed, overhead kitchen lights.

As he walked through the dark dining room to his master bedroom, he bumped into a chair. It scraped loudly on the wooden floor. "Dammit!"

"Liko, is that you?"

Light from the street filtered through the sheer curtains behind the couch where Charla now sat up, resting on an elbow. He saw her silhouette.

"Yes," he said. "It's me." Liko stood in place, unsure how Charla would react to him.

She lay back down, again resting her head on a cushion.

"Charla, I am so sorry. What I said was mean, cruel, and unkind. I am so very sorry. What I said was not the way I feel about you. Not at all."

He walked across the room and knelt beside the couch, placing his hand gently on her shoulder.

She looked sideways and up at him.

He could now see her face. She had removed her makeup and eyelashes. He thought she looked handsome. *You're a chameleon*, he thought. *Beautiful with makeup, handsome without.*

"I hope you can forgive me," he said. "I was a real asshole."

"Oh, Liko," Charla said. "I forgive you."

"I tried to apologize last night but..."

"I know." There was sadness in her voice. "I was on the porch and the door was cracked open. I heard your apology." She placed her hand on his and squeezed it affectionately. "I was just too upset to answer."

"Do you want me to get you a blanket? Do you need anything? A glass of water?"

"No, no, I'm OK."

She squeezed his hand and held it tight. "Liko, you can be so kind and gentle. I wish more people were like you."

She sat up and said, "Please, sit here beside me." She patted the couch. "Are *you* OK?"

"Yes. I'm fine."

"I saw you on CNN reading the press release. Well done. And I saw Augustus's thugs attack the reporters."

"I stayed out of it." He half-smiled, remembering how the FBI commando had held him back. "I think the thugs will be identified and warrants issued for their arrest. That's what members of the FBI swat team told me, anyway."

"And I hope the reporters and protestors sue his ass."

"With his money, I doubt he cares."

"Before I forget, Danielle called." Charla smiled. "She was worried about you. She asked you to call when you got home."

"Thanks," Liko said. "I imagine it's too late now."

"She sounded worried. I think she saw the clash on CNN, too."

"I'll give her a call." Liko sniffed the air. "Hey, something smells wonderful."

"Trader Joe's Pumpkin Bread."

"Oh my god!"

"I baked it earlier. It's on the kitchen counter wrapped in foil. Help yourself."

The relief Liko felt was palpable. He looked at Charla and he knew why. He knew that he was lucky to have her as a friend.

"If you will excuse me, I'm going to call Danielle, and then I'm going to eat half a loaf of pumpkin bread."

When Liko called Danielle, they talked about the clash between Augustus' thugs and the peaceful protestors at the coal plant. As the conversation wound down, Danielle said, "I've decided to don my reporter's vest and do a profile on Marimichael. Are you interested?"

"Absolutely," Liko said without hesitating.

"I usually work alone," she said. "But you're interested in Marimichael, and I could use the company. I plan on driving to West Virginia tomorrow to meet one of her mentors, a man she told me about. Someone she admired. I could use a driver and companion. Still interested?"

"Oh yes!" Liko could hardly contain the enthusiasm in his voice. "When do we leave?"

16

Young Environmentali st

The ornithologist lived in isolation in the countryside, choosing to live close to nature as Henry David Thoreau had once lived, deliberately. Of course, his neighbors did not understand him and they mistook him for a hermit. They were unaware of his doctorate from Cornell, or the more than one hundred scientific articles he had written about birds. He insisted that Liko and Danielle address him by his nickname, Ace.

Ace wore a bark-colored Barbour Ashby jacket, unzipped in front because the sun was bright, the sky light blue, and the temperature had crept into the mid-fifties. His loose-fitting denim jeans were held up by a bridle leather belt. His leather hiking boots were scuffed and well worn.

Danielle was dressed in a full-length wool coat. She had a lanyard around her neck and an identification card with the word PRESS above her photo. The photo was a younger Danielle, before her hair turned gray.

A digital voice recorder rested on her chest, hanging on a black cotton lanyard next to her photo ID. She was also armed with a reporter notebook and pen.

Liko was wearing a bright orange, quilted, goose down hoodie. That was one of the quirks about global warming: it was hard to predict the weather extremes. Every month set a record high temperature somewhere, and local temperatures often swung from low to high, and high to low. It was now an unseasonably warm December day in West Virginia.

Rhododendron bushes grew wild to the back door of Ace's house. In the distance lay a tilled-under vegetable field and then a large pond. Admiring the view, Ace, Liko, and Danielle sat together at a large picnic table, sipping mint tea from three artisan mugs, smelling the fresh sprigs of mint Ace had just picked for them.

"You're an ornithologist?" Danielle asked. "A bird man?"

"My profession and hobby," Ace answered.

"Your feeder is doing well." She tipped her head toward a large bird feeder mounted at eye level atop a sturdy wooden post. A northern cardinal flew from the twisted branches of a rhododendron to the cedar-wood feeder, scaring away a tufted titmouse and scattering birdseed from the feeder to the ground. Beneath the feeder, finches and sparrows and black-capped chickadees picked through seeds that rained down on them. "Any tips for beginners?"

"Start with a platform feeder and place it high off the ground."

"Any other tips?" she asked, pitching another softball question.

"Set out several birdhouses. And a birdbath, too."

They watched the cardinal dominate the feeder, chasing away the other birds. "The winter redbird is our state bird," Ace volunteered. "The males are bright red. The females less colorful."

Liko had never seen a bird feeder before. Or a redbird. He wondered if he could set up a feeder on the sixth-floor balcony of his apartment. *Would the birds find it?* he wondered.

"Any other suggestions?" Danielle asked.

"Keep the birds safe from feral cats." With that piece of advice, the tone of the interview changed. "You know," Ace said, "I follow your blog."

Liko could see the surprise on Danielle's face.

"You wrote a piece debunking sustainable growth. That took a lot of courage."

"You surprise me, Ace."

"No, you surprised me. It took a lot of courage to post that article. You wrote the facts, the truth. You nailed the problems with sustainable growth: finite resources versus human population growth and human nature. I've been a fan of yours ever since."

"So you know why I'm here?"

"Yes. It was Marimichael who told me about your blog."

"Well then." Danielle turned on the recorder. "Let's start, shall we?" She set the recorder in the middle of the table, between herself and Ace. They were seated across from each other. Liko was seated on Danielle's right with a clear view all the way to the deciduous trees that grew along both sides of the creek bed in the far distance, beyond the garden and the pond.

"Fire away."

"You know the O'Briens?"

"Yes, I know the family well." Ace removed his field hat and ran his fingers through his thin gray hair. "It's tragic, the losses the O'Brien's suffered—loss of health, life, and property—and all because of Augustus Dimer's coal company."

"Tell me about the O'Briens."

Ace put his hat back on. "Their father was a union man and a mine worker. He died of black lung disease when Marimichael was in grade school, and Fintan and Shane were in junior high. I took them under my wing. Marimichael used to visit after school. Her brothers, too."

From his earlier meeting with Danielle, Liko already knew that Marimichael became an activist shortly after her father died of black lung disease. He now listened intently.

"You still in touch with them?" Danielle asked.

"I heard they started a revolution." Ace smiled. "Imagine that: three kids from the holler starting a world-wide revolution."

"When did you last see them?" Danielle asked.

"Marimichael? Not too long ago."

"Her brothers?"

"About the same." He smiled and added, "Shane told me to expect a visit from the 'authorities.'"

Liko and Danielle glanced at each other.

"When was this?"

"About two weeks now."

"What were their parents like?" Danielle asked.

"Their father, Jacob, he had a profound influence on the boys. He loved the outdoors and hunting. Shane and Fintan are expert outdoorsmen. Marimichael, too. She shared her father's love of nature." He paused to sip his mint tea, and smiled at Danielle and Liko.

"She's special: she has always spoken truth to authority. Even as a child. She got expelled from grade school for asking too many questions. They said she was disrespectful, but I suspect she frightened her teachers. Showed them how ignorant they were.

"She no sooner learned how to spell 'environmentalist' than she became one. Joined a local group protesting a county road that would cut through a wetland. She's the youngest environmentalist ever arrested, at least in West Virginia.

"The sheriff locked her up with a bunch of men—and overnight, too. He said that he mistook her for a man because when he arrested her, she cussed him out with such foul language that he never suspected she was a girl.

"She hired herself a lawyer. Sued the sheriff and the county. Imagine that: a teenage girl suing the county sheriff. She had rare confidence and grit." Ace paused and smiled.

"As part of the settlement, the county agreed to leave the wetlands alone. They re-routed the road. It saved a lot of birds. And a few threatened frogs, too.

"I helped her start a local birding group. She was charismatic and soon had a following. Lots of local boys," he chuckled. "After she graduated high school, she became a member of a national group of environmentalists, but then she had a falling out with them. They were moving too slow for her. Some folks talk, and some folks do.

"She told me that she wanted to take on the owner of the West Virginia coal companies. She was determined to find a way to do it." Danielle lifted her pen for a moment and shook out her hand.

"She attended West Virginia University in Morgantown. She surprised everyone, though, when she joined the military, following in the footsteps of her brothers. Shane had joined the army, Fintan the marines. Marimichael enlisted in the army. That's where she learned discipline and leadership. I imagine you've seen her tattoos?"

"The sharpshooter tattoo?" Liko asked.

"Yes," he said. "She likes to illustrate herself with her accomplishments. Have you seen the war meteor?"

"Yes," Liko said.

"It's from a painting by Frederic Church. *The Meteor of 1860.*"

Liko recalled the tattoo: a giant meteor scorching across her back, brilliant and unlike anything he had ever seen.

"What else can you tell us about her?" Danielle asked.

"When she left the service she moved into Shane's home near Harpers Ferry. She became fascinated with John Brown's legacy."

"Why John Brown?"

"He was a man of action. He lived his beliefs."

"Anything else?"

"Marimichael attracts loyalty." Ace looked at Liko, intently. "And she is intensely loyal."

Ace turned his attention back to Danielle. "May I ask you a question?"

"Yes."

Ace grinned at Danielle. "For the record?"

She laughed. "Of course."

"What's your take on the climate crisis?"

There was a long pause as Danielle considered her answer. She sighed and then said, "I think the climate crisis will continue no matter what we do. Net zero greenhouse gas emissions by 2050? Never happen. We already wasted our opportunities. It doesn't really matter what we do now. More monster hurricanes. More killer heat waves. I look ahead to 2100 and 2200 and see the inevitable decline of the human race. Most species will be extinct. Only caged animals will survive."

"And you, Liko?" Ace asked.

Liko looked at Ace and Danielle. "At first I was hopeful, especially when I saw the public pulling together during the coronavirus pandemic. That gave me hope. I thought, now everyone understands the importance of flattening the curve, slowing the growth of the invisible enemy, whether it was a coronavirus or greenhouse gases.

"But we learned nothing. The airlines, cruise ships, and the tourist industry all resumed business as usual. Even today there are more planes flying, more ships sailing, more cars on the roads. And few companies continued to video conference or allow employees to work from home. Humans once again buried their heads in the sand. Long story short: Giving the current state of affairs, I think that it is now too late."

Ace and Danielle both stared at their feet, listening to his stark assessment.

Liko added, "Do I feel guilty? Yeah. Do I feel responsible? Yeah."

"You have a nice place here," Danielle said, trying hard to change the subject.

"Thank you."

Liko looked out across the vegetable field to the three- or four-acre pond. "Is your pond stocked with fish?"

"Yes," Ace answered. "It provides me with most of the protein I need—catfish and brim and occasionally a large bass."

"Is she a good swimmer?" Liko asked.

"Marimichael? Yes." Ace smiled at Liko. "She learned right here in my pond."

"Have you lived here long?" Danielle asked.

"I moved here after my wife died. We used to live across from the Dimer Coal Plant. Marimichael's father and I were neighbors there."

Danielle continued to question Ace for several more minutes. Occasionally something in his yard distracted him: a squirrel, a chipmunk, or a fish splashing in his pond. In the middle of one question and answer, he paused and said, "Now that's a pretty bird."

It was a chunky bird with gray head, black bib, and yellow chest. The bird opened its bill and warbled a high-pitched screechy note.

"What is it?" Liko asked.

"A mourning warbler," Ace answered. "It's one of the few remaining songbirds. Did you know that billions of songbirds have vanished?"

"No," Liko said. "Is it really that bad?"

"Yes," Ace answered. "Bird populations in North America have declined by thirty percent since 1970."

Now it was Liko's turn to stare at his feet in dismay.

After Danielle finished questioning Ace about the O'Briens, Ace turned to Liko. "Young man, would you like to see my vegetable garden?"

"Yes, I—"

"Sorry," Danielle interrupted. "We need to go."

"You city folk are always in a hurry," Ace said, politely admonishing both Danielle and Liko. "I hope you can come back and have a longer visit. You can fish or walk in the fields or along the creek."

Liko spoke up. "I'd like that."

"You can borrow one of my rod and reels."

"How about bait?"

"There are plenty of worms under the rhododendrons." Ace nodded his head towards the thick mat of decaying leaves.

Almost as an afterthought, Ace said, "Are you driving past Harpers Ferry?"

"Yes," Danielle answered.

"You should check out the museum."

"Oh, I've been there before. Seen the museums." Danielle pulled up the collar of her wool coat around her neck. The wind had suddenly picked up. "John Brown doesn't interest me anymore."

"But Marimichael does?" Ace said.

"Yes."

"Well then, you should stop in Harpers Ferry. The O'Brien's opened a gift shop there. They converted a small room in the back of their gift shop into a little museum." Ace smiled. "I think it'll interest you."

After he gave Danielle the details about the O'Brien's museum, she said, "Thank you for agreeing to the interview."

"You're welcome. I look forward to reading your article."

"I just thought of one more question. Can we go back on the record?"

He looked at Danielle and said, "Yes."

She turned on the digital recorder and asked, "Actually,

I have two questions. What is Marimichael's role in the rebellion? And do you think the rebellion will succeed? Or how will this all end?"

Liko smiled. She had asked three, not two.

"What role is Marimichael playing?" Ace seemed to think carefully before he spoke. "She has always been a leader. Never a follower."

"How will all this end?" Again he paused to think before he spoke. He looked out across his vegetable field to his pond. In the distance a fish jumped. It was airborne for a moment, a brief moment, and then it fell back into the water. The splash was loud enough that Liko, Danielle, and Ace heard it, even though they were a great distance removed. They watched the ripples spread across the surface and strike the unyielding shoreline. They dissipated before they reached the far side of the pond.

When Ace finally answered he said, "You should visit the O'Brien's museum in Harpers Ferry."

"What's Harpers Ferry?" Liko asked Danielle as they pulled out of Ace's long gravel driveway and back onto the county black top.

"Harpers Ferry? It's where the Potomac and Shenandoah Rivers meet. It's also where John Brown tried to start a slavery revolt just before the Civil War. He seized an armory where the federal government made pistols and rifles. He wanted to arm the slaves."

As Liko gazed at the road ahead of them, a vision of John Brown came to him. And then he thought about the Civil War. Was violence the only way to overcome the powerful forces that kept men enslaved?

And now, were the O'Briens right? Was violence the

only way to overcome the powerful forces that profited from burning fossil fuels?

17

Harpers Ferry

Liko had not been in a private museum since he lived with his great-aunt in Hawaii. She had a marvelous collection of Hawaiian artifacts in her home that she kept on display. Liko had learned about his Hawaiian heritage while browsing her collection. She kept a Hawaiian warrior mask and koa-wood warrior club safely locked behind glass in a cabinet in her living room. The warrior bowl was embedded with the teeth of victims.

Liko now found himself entering the O'Brien's public museum in the back of their gift shop. Admission was free. There were no crowds, no reporters, no heightened interest in the store, all of which surprised him and Danielle. Perhaps it was because the name of the store was John Brown Gifts & Legacy, and there was no apparent reference to the owners: Fintan, who had assassinated the President of the United States, and Shane, who had taken an oil baron's wife and grandchildren hostage. Marimichael's name also was not advertised.

The gift shop will soon be infamous, Liko thought. He paused to read the sign above the door that separated the shop from the museum: THE LEGACY OF JOHN BROWN AND CLIMATE REVOLUTION. At the entrance was a portrait of John Brown, his hair messed up, his eyes wild and crazy. Liko walked through the door and strolled to the center of the one-room museum, turning to face an old stone fireplace.

He read the nameplate on the painting hung above the mantel: "Frederic Church. Meteor of 1860." He recognized the painting as the beautiful tattoo on Marimichael's upper back.

He took two steps backward and admired the giant meteor scorching across the sky, the iron-stone exploding into two large and many small fireballs, each a new meteor, each hot white, and each shedding a tail of dust that burned yellow-orange-red. Church had painted the Earth-grazing meteor passing through the earth's atmosphere. *Did it strike the earth? Or did it pass through our atmosphere, just missing us?*

"Siri," Liko said, "provide information about 'The Meteor of 1860.'"

Siri instantly responded: "John Brown, Abolitionist: The Man Who Killed Slavery, Sparked the Civil War, and Seeded Civil Rights, by David S. Reynolds, page 383: 'a tremendous meteor that raced in a bright streak across the morning sky and then exploded. The meteor was seen as far north as Albany, New York, and as far south as Fredericksburg, Virginia. When the meteor burst, it gave off what one journalist called "a series of terrific explosions, which were compared to the discharge of a thousand cannon.""""

Danielle stepped forward to browse three books stacked on their sides, one atop the other, on the mantel. "A bible... a *Summary of Oliver Cromwell's Campaigns*, and ... *Les Mis*, by Victor Hugo."

Liko had seen the musical *Les Misérables*, so he was familiar with the ex-convict Jean Valjean and his struggles for redemption during the working-class revolt in ninetieth century France. He had never heard of Oliver Cromwell, though.

"May I see the bible?" he asked.

Danielle handed the old book with its creased black cover to Liko.

He perused the inside cover and the first few pages, stopping when he discovered a handwritten genealogical tree. He worked his way through the tree until he found the names of Shane, Fintan, and Marimichael. Their names had been neatly written in cursive in black ink. "It's the O'Briens' family bible."

Liko flipped through the pages, stopping when he came to a passage underlined in red ink. He read aloud: "Leviticus 25:44, 'You may purchase male or female slaves from among the foreigners who live among you.'" He looked at Danielle. She was listening. "Leviticus 25:45, 'You may treat them as your property, passing them on to your children as a permanent inheritance.'"

"The bible justifies slavery," Danielle stated, matter-of-factly. "Why aren't I surprised?"

Liko flipped through the pages, stopping at the red-underlined passages. He read Exodus 21:2-6 silently to himself, and then he read Leviticus 21:7, which was also underlined in red. He shook his head and said, "The bible allowed the separation of husbands from their wives and children. And it allowed men to sell their daughters."

"Nomadic, desert tribes. A different time," Danielle commented. "A different reality."

"The New Testament is just as bad," Liko said. "Listen to this: Ephesians 6:5 — 'Slaves, obey your earthly masters with deep respect and fear. Serve them sincerely as you would serve Christ.'"

"One more verse," he said, "Titus 2:9 -10, which says: 'Teach slaves to be subject to their masters in everything, to try to please them, not to talk back to them, and not to steal from them, but to show that they can be fully trusted.'"

"I challenge you to find a passage condemning slavery," Danielle said.

"The eleventh commandment," Liko quipped.

"The what?"

"The eleventh commandment," he reiterated. "You know, 'Thou shalt not enslave thy fellow man.'"

They would have laughed if it weren't so tragic.

Liko returned the bible to the mantel. He moved to the right of the fireplace and stood beside Danielle who was now looking at three framed photos, side by side. He read aloud the nameplate beneath the first photo. "Mary Harris Jones. Miner's Angel."

He read the title in the legend of the next framed picture, "Blair Mountain. USGS Topographic Map." Liko stared at the brown, squiggly, contour lines on the light green map. "Why frame a map of Blair Mountain?"

"The Battle of Blair Mountain?" Danielle answered. "Every West Virginian knows that story. It's part of their history, their blood."

Liko shrugged his shoulders.

"It was the largest insurrection in the United States, after the Civil War," she said. "A full-scale war fought along a 25-mile front in West Virginia. Eight thousand armed miners seized trains and fought mine guards, police, and the state militia." She sighed and continued, "Of course they lost."

"Why were they fighting?" Liko asked.

"They thought democracy had failed them," Danielle said in a subdued tone of voice.

Liko read the nameplate beneath the last picture: Frank Keeney. Turning to Danielle, he asked, "Who were Mary Jones and Frank Keeney?"

"Mother Jones was an Irish-born union organizer," Danielle said. "During her lifetime, she was known as the most dangerous woman in America. She recruited Keeney. He led the miner's rebellion."

They proceeded to the next photo: Abraham Lincoln. Below the photo and inside the frame was the quote: "As I would not be a slave, so I would not be a master."

"The Golden Rule," Liko said.

Next to Lincoln was a framed newspaper article. It was yellow and faded and Liko wondered if it was an original. The date was 1907, and the headlines read: "425 Are Dead" and "No One Ever Prosecuted" and "We Have to Protect Ourselves."

"What is this all about?" Liko asked, nodding his head towards the article.

"A tragedy," Danielle said. "A coal car broke loose and barreled into a mine. The explosion killed all the men, instantly. I heard a story," Danielle added. "One man was seated, eating a sandwich. They found him dead, with his lunch bucket on the ground between his legs and the sandwich still in his hand."

"Shit!" Liko said, an incredulous smile filling his face. He looked at Danielle. She was smiling.

"It wasn't funny at the time," she said.

On the next wall, they came face to face with a polar bear, rhinoceros, and an emperor penguin. The images were large, but not life size; the room was too small to display life-size photos of the large animals. Nevertheless they looked real. Liko read the placard to the left of the photos: "Getty Images by Cody Long."

A professional photographer? Liko wondered.

He recalled his dinner conversation with Shane and Marimichael. She had asked him for his opinion about the sixth extinction, the man-made extinction of most life on earth. He now recalled how she had looked at him expectantly. The expression on her face had been hopeful — hopeful that he understood. *And my answer?*

The eye of the rhino was in sharp focus and drew him in. Liko wanted to step into the picture and jump on the back of the massive mammal. Was it a white rhino or a black rhino? Liko suddenly realized he didn't know the difference between the species. Wasn't one more threatened than the other? Or was one of them already extinct? Suddenly realizing his ignorance, he felt embarrassed. *Did I see a rhino in the Honolulu Zoo,* he wondered, *when I visited Uncle Keahi?*

His attention shifted to the polar bear. It looked famished and tired, sitting alone on a small piece of floating ice. Liko had seen similar photos before, but this was a cub. It had no chance of survival. *And neither does its species,* Liko thought. *Except in a zoo.*

Liko barely glanced at the emperor penguin. He felt like he had to catch up with Danielle. She had moved on to the third wall.

Marimichael appeared before them in a field of wild blue flowers. Her photo was life size. She was on her knees, her arms outstretched, her hands cupped as if offering museum visitors a gift. In her palms rested a small butterfly. Her face beamed with joy. Liko read the placard: "Blue Lupine (*Lupinus perennis*) and Endangered Karner Blue Butterfly (*Lycaeides Melissa samuelis*)."

He gazed at the Karner Blue. *Breathtaking.* Deep violet-blue wings outlined in black and then fringed in white. He inhaled sharply.

Liko recognized the flowers and the butterfly as the tattoo on Marimichael's shoulder. *This is her passion, her motivation, to stop the mass extinction of life on earth.*

As they exited the small, one-room museum, the gift shop attendant asked, "Did you enjoy your visit?" The attendant was young, probably a student. Her brown hair was pulled back in a ponytail and she had an infectious smile.

"Yes," Danielle answered. "Do you know the woman in the photo, kneeling in the flowers, holding the rare butterfly in the palm of her hand?"

"No," the attendant answered. "I just started work this month."

"I see," Danielle said. "Her name is Marimichael. I believe she is your employer, or at least one of them." Danielle handed her a business card. "Please ask her to call me, OK?

"I'm sorry. I don't know her." The attendant gave each of them a thumb drive and a brochure. The thumb drive had a wooden case and was stamped with an owl icon. A small sticker affixed to the case read "Join us!"

Liko turned to Danielle and held out the thumb drive.

"Marimichael gave me one of these after your lecture."
And I thought it was something special.

They stepped out of the O'Briens' gift shop into the bright sunlight.

"You know that John Brown was a murderer, don't you?"

Liko knew nothing about John Brown, so he shrugged.

"In Kansas, he dragged five men out of their homes. Shot one in the head and watched as the other four were hacked to death with broadswords. He was a murderer, a liar, and a terrorist. After he kidnapped some citizens here in Harpers Ferry, the townspeople rose up against him."

Opening her car door, Danielle continued, "Shane and his brother call themselves climate change revolutionaries, but they're just domestic terrorists. Like Timothy McVeigh, or the Protestant Minister Paul Hill, or that mail-bomber nut, Ted Kaczynski."

They looked at each other and their unasked question hung in the air: "Is Marimichael a terrorist too, like her brothers?"

On the drive from Harpers Ferry to DC, Liko and Danielle talked about Shane and his attempt to blackmail Augustus Dimer.

"Do you think Shane's revolution is justified, given the gridlock in government today?"

"Justified? No. But I have no doubt that he believes his cause is just and his methods necessary. The fossil fuel industry does have a stranglehold on the Senate."

"Have you always wanted to be a reporter?" Liko asked.

"Yes. I've always been curious about things. Like you, I am a seeker." She looked across the front seat at Liko and smiled. "It gives me a lot of freedom. And it empowers me to ask anyone anything."

Liko nodded. "Did you ever report on something that you didn't want to?"

"Yes. Sometimes people get hurt. Sometimes people don't want to know the facts, especially when the facts contradict what they think or believe. But I have always reported the truth."

"Has anyone tried to hurt you?"

"Face to face, no." Danielle focused on the road ahead. "But sooner or later it happens to all good reporters. It comes with the job."

"What happened?"

"I've been doxed, swatted, and dog-piled."

"What?" Liko had no idea what she had just said.

"Doxing is when someone gets your private information, like your phone number or address, and they publish it online. It's dangerous, because when you publish a story that makes someone angry, and then someone gives them your address, it can be life-threatening. It happened to me. Someone killed my dogs right in my back yard."

"Swatting is when they tell the authorities lies about you and then the police show up with a search warrant, or they come to arrest you. It's not funny when the FBI or the IRS knock on your door. The IRS visited me. I know a journalist who had a SWAT team crash through his front door at two o'clock in the morning."

"I can guess what dog-piling means," Liko said.

"I imagine you can. And believe me, it ain't fun when EVERYONE is jumping on you."

"So what did you do?"

"I paid a company to remove my personal information

from the internet. It wasn't cheap, and I have to pay them every year, but believe me, it has been worth it. You can do it yourself, but it takes a lot of time. I used DeleteMe before I heard the private company."

She looked at Liko. "I did it to protect my parents. There are people out there who are crazy and ruthless."

Liko knew that was true. "How about social media?"

"What about it?"

"Do you use Facebook and Twitter and Instagram?"

Danielle barked a loud laugh. "I've deleted all the content and personal information on my Facebook and Twitter accounts." She smiled. "I kept the accounts, though, so I can follow and look at other people's information." She laughed. "But if you visit my site, all you'll see is a blank page.

"How about you?" she asked.

"I don't use social media. Never have."

"Then don't start. Twitter became an authoritarian platform for Trump and now the Republicans. The company knows that Trump's a liar, but they refuse to reign in his 'conways.' You know what a 'conway' is, don't you?"

"I can guess." Liko surmised that she was talking about Kellyanne Elizabeth Conway, a counselor to President Trump, who popularized alternative facts. Conway was a political, Orwellian spin-master.

"And don't use Facebook," Danielle said, continuing her rant. "After Cambridge Analytica shared our personal Facebook data with the Trumpers, who would ever use Facebook again?"

Liko nodded his understanding.

At the next gas station, he filled the tank and they traded seats. Once in the passenger seat, he almost immediately fell asleep. He woke up with a loud snort. *Embarrassing.*

"Rested?" Danielle asked.

"A little," he said. "How long was I asleep?"

"An hour."

He discovered that the radio was now on. "Any news?"

Danielle laughed. "No. I tried to find NPR, but no luck."

Liko was becoming use to her laughter. He was starting to enjoy it, even though it was often a derisory, ridiculing laugh.

"Where is Edward R. Murrow when we need him?"

"Who?" Liko asked.

"You've never heard of Edward R. Murrow?"

"Nope."

Danielle shook her head. "The golden age of radio? Reporting from London during WWII as it was bombed by the Nazis? No? Well, I'm not surprised. Nowadays, one gets nothing but an occasional rip-and-read story on the radio."

Liko yawned.

"Ever folded a newspaper?" she asked.

"What?" He smiled. "Of course." But then he confessed: "A time or two."

She smiled. "Well, I imagine that's more than most Millennials."

I'm not a kid, Liko thought. He looked at her. *She knows that. Maybe I should say, "OK boomer."* The idea made him smile.

It was true, though; he rarely read a newspaper. "They used to have newspapers in Starbucks. I'd glance at the headlines while I was ordering a coffee, but I never bought one. I don't think they carry newspapers anymore, do they?"

"No. I don't think so."

"I subscribe digitally, though," Liko said. "To *The Washington Post* and *The New York Times*."

"Two papers? I'm impressed."

"Yep." Liko felt a bit proud of himself. "How about you?"

"I subscribe to the *Post* and the *New York Times*. Also *The Guardian* – that's a British paper with a great environmental staff. Occasionally I read *Financial Times* and I donate monthly to *InsideClimate News*."

"Impressive," Liko said. "The *Post* is good. I like their videos. And they have ..."—Liko tried to think of the word – "interactive articles."

"Infographics?"

"Yes. Infographics. I thought their infographic on '2 Degrees C: Beyond the Limit' was exceptional."

"It was," Danielle said, nodding her agreement.

"The *Post* was my go-to site for information during the coronavirus epidemic."

"They're good," Danielle turned off the radio. "But they're not perfect. Always insist on facts, and science."

"Not perfect?"

"They totally blew the weapons of mass destruction and supported Bush's war in Iraq."

Liko nodded. "I didn't subscribe back then."

"They were pro-war. Bob Woodward went on the *Larry King Live* show and said there was zero chance we wouldn't find weapons of mass destruction. Sadly, he was one hundred percent wrong. The *Post* opinion editor was pro-war, too." Her expression was droll. "Always insist on facts and evidence."

Liko suddenly realized that her bark of laughter often carried feelings of intense pain, too.

"Even the *Post* opinion editor said we should invade."

Liko felt tired. "You know," he said, "it's becoming harder and harder to remain an optimist in such pessimistic times."

"Well," she said. "The *Post* did own up to their mistake. They published an apology. It was a piece by Howard Kurtz titled 'The *Post* and WMDs: An Inside Story.'" She looked at Liko. "I recommend you read it."

"Sure thing," he said, shaking his head in dismay. "I'll add it to my reading list. My ever-growing reading list."

They were quiet for a while as they continued their drive towards DC.

"Anything you'd like to know about newspapers? Reporting? The reporter's life?" She glanced at Liko. "Just ask."

He mulled it over for a moment but decided he could Google anything that interested him.

A few more miles down the road, she asked, "You know all about Daniel Ellsberg and the Pentagon Papers? Woodward and Bernstein, and Watergate? Edward Snowden and the NSA?"

"Yes, ditto, ditto."

She smiled. "The *Post* won Pulitzers for those stories."

"Would you like to win a Pulitzer?" Liko asked.

She barked, yet again. "Sure."

"But can you? I mean, you're not with a newspaper. You write a blog."

"Actually, yes." She smiled. "David Wood was the first to win a Pulitzer for a blog. He wrote a series on wounded veterans."

"Well, maybe your piece on Marimichael will win?"

Danielle laughed so loud and long that she had to grip the steering wheel with both hands.

That evening, Liko dropped Danielle off at her home and then returned to his condominium building. He clicked the button on the fob and the parking garage door opened. He drove to the lowest parking level, the fourth floor, and pulled Zahi's Lexus 350 into his assigned parking place. Liko had still not gotten use to owning the condominium, the car, and all of Zahi's stuff.

He used the fob again to call down the elevator. He felt out of place as he rode to the sixth floor of the high-end building.

Once again Charla had waited up for him. This time, though, she was awake, seated on the couch, and watching MSNBC.

"Hi," he said. "You're up late." *As usual.*

He sat down beside her and leaned his shoulder against hers. They were quiet for a minute. He was tired.

"Want some popcorn?" she asked.

"No thanks. I'm too tired to floss."

They were quiet for another full minute. They both felt comfortable with the silence. "I'm lucky to have you as a friend," he said.

"Me, too," Charla agreed.

18

Raid and Manifesto

Liko awoke to a phone call from Danielle. "Shane is dead. So is Heidi Dimer. Turn on CNN."

Liko drew a quick and shallow breath. He glanced at his iWatch: 6am.

He rose from his bed and went directly into the living room. He turned on his television and switched the channel to the breaking news: "Shane O'Brien, the brother of Fintan O'Brien who assassinated the president, is dead, along with two other terrorists."

The film crew highlighted flame and smoke rolling from the mansion, and then they broke away to the chaos on the front lawn. An ambulance, fire engines, police cars and officers, and para-military forces with men dressed in solid-black uniforms were all crowding the scene. The camera cut to reporters massed behind a police barricade.

Liko held his breath for a moment as he switched to MSNBC. "Sources have confirmed that Heidi Dimer, the wife of Augustus Dimer, is dead." Stunned, Liko watched for another minute. When the station failed to provide further information, he switched to FOX news.

"Thank GOD my grandchildren are all right." Augustus cried into a microphone held by a FOX News reporter. "But my poor wife." Tears streamed down his cheeks.

Tears of joy. Liko clenched his jaw and closed his fists. He had no doubt that Augustus had a mistress, and that he would soon start a second family.

FOX News appeared to have a monopoly on airtime with Augustus, but their reporting was lightweight. Liko felt nauseated knowing that Augustus was getting away with murder, literally. The reporter asked none of the hard-hitting questions Liko expected, so he cussed out the inept reporter and switched stations to the BBC.

In a thick British accent, a BBC reporter said: "The mercenaries were employees of KUDZU4 LLC., a private military company that specializes in providing personal protection for financial institutions, banks, and fossil fuel utilities. KUDZU4 LLC is notorious for its operations in Yemen and other Middle Eastern countries."

The BBC reporter said that KUDZU4 LLC owned training camps in the southern United States. The camera cut to footage of a large camp in the pine forests of Alabama. In the background was a shooting range. The camera cut again, but this time to research facilities in Huntsville where the company designed drones and electronic monitoring devices. According to the reporter, the average salary of KUDZU4 LLC commandos exceeded $1500 per day, making them the most highly paid mercenaries in the world.

Liko slumped back into his Italia chair. He covered his face with his hands and closed his eyes. Heidi had predicted her own fate. *He's a murderer,* Liko thought. *Ruthless and immoral.*

The eco-terrorist response was swift. Before the firemen had contained the blaze at the mansion, the terrorists posted a video on YouTube. Liko was seated at the bar in the Grillfish with Charla, waiting for a lunch table. The Grillfish was in the block north of Liko's condominium.

When Liko saw Marimichael appear on the television above the bar, he took in a sharp breath. Charla sat up straight on her bar stool and stared at Marimichael, too.

"My name is Marimichael. My brother Fintan assassinated the President of the United States." She paused as someone off-screen handed her a rifle. She struggled for a moment with the weight of the gun as she laid it across her lap.

Liko felt his heart kick backwards. The woman he had dinner with was seated on a floor mat, dressed in casual street clothes and Nike cross trainers, and she had a rifle in her lap.

A man farther down the bar turned to his neighbor seated next to him and said, "That's a Crazy Horse Rifle."

"Army issue?"

"Could be. It has the black phosphate finish."

Resting her hands on the marksman rifle, Marimichael said, "My brother Shane died this morning. An oil titan paid mercenaries to assault his mansion, even though his wife and grandchildren were inside." Marimichael spoke clearly and calmly, even though her voice was filled with anger and sadness.

Liko felt mesmerized. He wondered if other viewers felt

the same. He glanced around the bar. Everyone was riveted to the television.

"We appeal to all grandmothers and grandfathers. End the stranglehold of the fossil-fuel titans and the fossil fuel industry." Her voice sounded as if she were speaking from her heart.

"We also appeal to all mothers and fathers. Protect the future of your children.

"And to young women, we say, 'Stand up! Seize your rights!'

"We appeal to all children. Join us. The future is yours. Girls, we fight for you and your rights, so you can improve the future of life on our planet.

"We appeal to all the scientists. Thank you for warning us about the current climate crisis.

"And of course, we appeal to all teachers. Join us. Educate young women so they can have fewer, yet healthier children. Equip them to face the challenges of climate change so they can make the best choices for their families.

"And we appeal to all students, too. Join us." Marimichael now spoke with confidence.

"We appeal to all farmers. Our future is in your hands. And our future depends on how you farm your lands. Adopt regenerative agriculture and silvopasture."

Liko had no idea what that meant. What was regenerative agriculture? And silvopasture?

"We appeal to all workers, world-wide. Build the onshore wind turbines and solar farms and electric cars. Install the rooftop solar panels. Capture and recycle refrigerants before they escape from our air conditioning equipment. We need you!

"We appeal to people of all religions — Jews, Christians,

Muslims, Buddhists, Hindus, and other devotees — join us! Being an atheist, I personally appeal to fellow atheists, too. Join us. We are all needed.

"We appeal to all governments and government workers to join us. Pass and enforce comprehensive climate crisis laws. Continue your commitment to protect our planet.

"Everyone! Everywhere! We appeal to YOU. Reduce your food waste. Eat a plant-rich diet!

"Today we see a revolution of the people, for the people, and by the people. Join us! Speak truth to authority!"

Liko liked that expression, "speak truth to authority." As he listened to Marimichael, he felt that she respected her audience and that she fearlessly spoke truth to them. He thought she embodied defiance.

"We appeal to all militaries of the world and all mercenaries." She picked up the M14SE semi-automatic sniper's rifle off her lap and leaned it against her side. "Do the honorable thing. Join us!"

"Look at the cheekpiece," the man at the bar said to his companion.

Liko looked at the light yellow-tan cheekpiece, too.

"It's worn," the companion commented. "Somebody has spent a lot of time at the firing range."

Liko thought his heart stopped.

"We have posted our manifesto," Marimichael said. "Many of you will not have the opportunity or the time to read it, so please allow me to summarize it for you." She adjusted the sharpshooter's rifle at her side.

"The abuse to our planet has been so violent that a violent response is now necessary. From this day forward, violence against the environment will be met with violence. It is now the only way to break the stranglehold of dark money and the fossil fuel industry." She looked directly into the camera and said, "Join us!"

The video ended. Liko looked across the top of the bar and regained his bearings as if awakening from a dream state. Charla was still seated next to him. The television had returned to an earlier scene at the mansion, highlighting the chaos that followed the mercenaries' assault on the residence.

The man at the bar said, "She's got a nice scope on that rifle."

Not wanting to be outdone, the other man said, "And a Vortex sound suppressor."

When the first man didn't respond right away, the second man added, "The bitch is a martyr-want-to-be."

"We have a table ready for you," the waitress told Liko and Charla.

Liko stood up and pushed his bar stool away, scraping the floor. *Marimichael, who the hell are you?*

19

Handguns, Assault Rifles, and Drones

Diamond Head grew larger as the drone flew towards Waikiki Beach. The weathered volcano appeared like a sleepy dragon that had pulled itself out of the ocean and collapsed onto the island.

When the drone reached the white beach, it flew parallel to the shoreline, first passing rows of pink umbrellas in front of the Royal Hawaiian Hotel, and finally stopping in front of a lone, tan umbrella that had been speared into the glistening sand. The drone hovered just offshore, and the onboard camera focused in-and-out on the man lounging in the shade of the umbrella.

The man was reading a sun- and water-worn copy of *The*

Economist — "The Climate Issue." He looked over the top of the magazine, annoyed by the humming of the drone. Another man, strolling down the beach in a skimpy blue swimsuit and a small towel slung over his shoulder, wearing sunglasses and a white baseball hat and leather sandals, looked in the direction of the drone, too. His path took him behind the sunbather.

The stroller shot the sunbather three times in the back of the head as he walked by: pop, pop, pop. He never broke his stride and he never looked back. He continued walking down the beach, his footsteps kicking up the hot, white sand.

The drone slowly turned one hundred and eighty degrees to face the ocean and the red sun. The sun appeared to be exploding as it set. The drone flew into the sunset until it ran out of power and dropped into the ocean.

Meanwhile, the assassin continued his stroll down the beach. The small Ruger LCP was barely visible in his large hand. It seemed like a toy, with its short, 2½-inch barrel. Nevertheless, he took the small towel and draped it over his hand and gun. When he reached the sea wall behind the Waikiki Aquarium, the assassin turned to face the setting sun. He threw the small gun as if it were a baseball into the green flash as the sun slipped behind the horizon and into the ocean.

When the assassin reached the far side of the aquarium, he turned towards Kapiolani Park. When he reached Kapiolani Boulevard he jumped into the back seat of a waiting Uber.

A drone hovered behind the fat man as he sat on his riding lawnmower, unaware that he was under surveillance. The lawnmower's Brigs and Stratton engine drowned out the humming of the drone.

When the obese rider reached the end of his long front yard, he noticed a stranger strolling down the sidewalk along the front of his property. The rider turned the loud, vibrating mower back uphill towards his Georgian-style house, paying no attention to the man who now passed on the concrete sidewalk behind him.

The stroller turned and settled into a shooter's stance. He shot the rider in the back of the head twice. Five more successive shots left a pattern of red wounds on the rider's broad back. He slumped over the small steering wheel. The lawnmower continued to climb up the hill towards the red brick house, still cutting grass.

The assassin tossed the Ruger LCP — a Coyote Special edition — into the newly mowed grass. It disappeared into the plush, dark green blades. The assassin walked 100 yards farther down the sidewalk and jumped into a waiting Lyft.

The drone flew ahead of the lawnmower and set down directly in its path. The image of the mower advancing and then the shiny metal blades whirling made dramatic footage. The drone was pulverized into tiny pieces that were sprayed across the width of the yard.

A drone provided a panoramic view of the baseball game, gradually zooming in on the spectators seated in the hometown bleachers. Black security guards sat on the front row. The drone focused on the white man seated on the second row, behind the guards. The man held a hot dog loaded with ketchup, mustard and glistening with onions in one hand and a Coke in the other. A paper plate of chips and baked beans rested on his lap.

The man took a bite of hot dog. He licked the yellow mustard just before it spilled out the end of the bun. He smiled and wiped his mouth.

Bang. The man's head snapped backwards. A second bang. The man's body blew off the bleacher and into a third row of white men. They panicked and shoved the man's body forward into the security guards who were still seated in the front row. When the security guards stood up, the man's body fell forward onto the ground. It bled out through a large hole in his back onto the baseball field. The plate of pork and beans had spilled onto the front of his white shirt, staining it brown.

A baseball flew towards the drone. A near miss. A bat whirled skyward and struck the drone, knocking it out of commission.

A drone hovered behind the two window cleaners who were standing on an outdoor platform with a bucket of soapy water, long-handled brushes, and squeegees. The building was a modern design, the windows curved and installed floor to ceiling.

Sunlight glistened off the tinted glass, which had been designed to reflect the sun's rays and to keep the interior of the building cool. The tinted windows, though, made it difficult for the drone to see into the building.

Both window cleaners bent at the waist and picked up their AR-15s, semi-automatic rifles with a green finish and fifteen-round capacity. They aimed their rifles towards the windows and the interior of the building.

They turned their heads and looked at each other and nodded agreement. The larger man turned back to face the window. He rapped hard on the glass. The window shimmered. They both tightly gripped the pistol grips on the AR-15s and fired.

As the windows shattered, the drone flew into the building and hovered near the ceiling. Its camera live-fed the chaos unfolding in the conference room. One man, later identified as the CEO of the fossil fuel company, jumped up and tried to hide behind a flip chart. Bullets pierced the paper and cardboard. Another man, later identified as a board director, pulled a facility meeting coordinator in front of himself for protection. Bullets passed through the coordinator and into the board director, whose body involuntarily danced against a white wall.

The drone remained near the ceiling, recording the carnage, then flew back out the shattered conference room window and turned to face the building and the assassins.

A wide-angle lens shot footage of the assassins BASE jumping off the platform to a grassy park across the street from the oil company's headquarters. The assassins left their rainbow parachutes adrift in the park and jumped into a nearby yellow taxi.

The AR-15s, their barrels still warm, rested on the platform outside the bloody conference room.

As he walked from a Baltimore, Maryland restaurant to a black SUV, the 7.62mm NATO caliber cartridge tore through Derichenko's body as if it were passing through warm butter. A bodyguard was walking beside him and two other bodyguards were waiting for him at the SUV.

The assassin calmly and carefully placed the Crazy Horse rifle into a dark brown gun case. The 856-yard shot had been made from a prone position with the trigger system adjusted to 2.5 pounds. Derichenko was dead before his body hit the hot asphalt.

The drones had taken video of each assassination and the rebels posted the footage on social media. Of course it was immediately taken down, but not before it went viral. The headline in *The Washington Post* read "IS ANYONE SAFE?"

20

Sports Grill and Bar

Liko was returning from the gym after a strenuous workout. The elevator door opened and he found himself face to face with Charla. She was dressed in a puff-sleeve zebra print top, wide mint-colored belt, and flowing yellow-green pants. And she wore wire rim glasses, which made no sense, since she had perfect eyesight—as far as Liko knew, anyway.

"Hi!" He was happy to run into her, happy to see her. "Where are you going?" He stepped out of the elevator.

"Out." She stepped into the elevator and pushed the button for the ground floor. She looked at him thoughtfully and then confessed, "I'm going to the bar."

"The bar?" He realized that she meant the Sport Grill and Bar, where bullies had harassed her because she was trans, and the bartender had ordered her to leave. "Why?"

"To confront the assholes."

He reached out and held the elevator door. "I'll go with you?"

"No."

He felt disappointed that she didn't want his help. He let go of the elevator door and it began to close. "Please be careful."

"I will."

And then he heard the elevator descending and she was gone. "Shit!" he muttered.

As he walked down the hall his mind raced. He thought about her confronting the bullies who had verbally assaulted her earlier in the week at the bar, when he had stayed home to watch a movie and she had gone alone to have drinks and to relax.

By the time Liko reached the front door of his condo, he had made up his mind: *I should be there just in case.* He turned around and quickly walked back to the elevator, but he had to wait a half minute for it to return. When he finally reached the lobby, Charla had gone.

The concierge stood up from his chair behind the front counter. "How are you tonight, Mr. Koholua?"

"I'm well," Liko answered. "And you?"

"Fine, Mr. Koholua."

Liko next exchanged "good evenings" with the doorman.

"Can I get you a taxi, Mr. Koholua?"

"No, I'm walking, but thank you," Liko answered. The bar was nearby, and he was sure Charla had walked.

He took off down the sidewalk at a fast pace, hoping to catch her. When he reached the Sports Grill and Bar, he climbed the stairs quickly, but then decided that he needed to slow down and get a grip on himself.

He saw her seated at the bar, alone. She was incredibly attractive. *Anyone disrespects you, I'll beat the shit out of 'em.*

She hadn't seen him, so he took a seat at a dark table in the back of an adjoining room, where he could see her but she would have difficulty seeing him. He ordered a pitcher of beer and waited.

Before he had finished his second glass of beer, a short, slender man sat down on the barstool next to Charla. He was dressed in a pink sweater and light sienna, twill trousers. He carried a brown leather satchel that he slipped from his shoulder and set on the ground at the foot of his stool. He placed a book on the bar counter, resting it between himself and Charla. When he turned, Liko saw his profile: Jack, the FBI man. His black hair was perfectly coiffed, held in place on the side with a large, decorative hairpin.

Well I'll be damned, Liko thought.

Liko looked at Jack carefully. His pink sweater was V-necked and was pulled over a white shirt. His navy tie boasted a flower motif. Jack looked like a model. *He dresses well,* Liko decided.

Charla noticed the book, pointed at it, and asked Jack something. Liko was too far away to hear their conversation, and he couldn't read the title of the book, but he knew that it had piqued Charla's interest. *Nice way to start a conversation,* Liko thought.

As Liko started his third beer, Charla and her 'date' were still talking and laughing. He noticed, too, that they were now sitting closer together. One or the other had scooted their bar stool closer. He had missed that moment, regretfully. He smiled.

After he finished the third glass of beer — and on an empty stomach — he decided that he should eat something. He ordered a hamburger deluxe with a side of sweet potato fries. When it arrived fifteen minutes later, Liko noted that the couple seemed to be doing well, so he shifted his attention to his meal.

It was a juicy, messy burger. He attacked it and it fell apart onto his plate. He needed three paper towels to wipe and clean his hands, leaving them still sticky with hamburger and barbecue sauce. He pushed the empty plate across the short, square table. His hands were a mess and he needed a restroom.

Scanning the bar for the men's room, Liko saw that two men and a woman now stood behind Charla and Jack. One of the men put his hand on Jack's shoulder and forcefully turned him.

That's assault, Liko thought.

Jack leaned backwards, instinctively, his upper body and head leaning out over the top of the wooden bar behind him, trying to get away. But he had no route of escape.

Liko sprang up from his seat. He bumped his table, sending an empty pitcher of beer sliding away from him. He contorted his body and caught the pitcher before it fell. He watched, though, as a half-full glass of beer slid off and broke. *Dammit!* He'd deal with it later. Setting the pitcher on the table he turned towards the front bar again.

Jack had regained his balance on the stool. He now sat upright and pushed the assailant back.

The man advanced again and Jack stiff-armed him, stopping him abruptly. With his free hand Jack reached for his glass of beer and poured it over the assailant's head. The man backed up a few paces, wiping off the beer as it ran down his face.

Yes, Liko thought. *That's what you have to do with bullies — push back hard!*

The second man now lurched towards Jack. Calmly, Jack removed a peineta from his coiffed hair and thrust it quickly underneath the man's chin. The five pointed teeth of the shiny brown comb pressed against the roof of the man's throat, stopping him abruptly. He stood on his toes, as if frozen, no longer daring to move.

"Don't!" A woman yelled.

The bartender, who had been watching from the other end of the bar, yelled, "Enough! This stops now or I call the police."

The first man, his head and shirt soaked with beer, moved towards Jack yet again.

"See that camera?" The bartender pointed to a camera in the corner of the ceiling. "You stop now or I call the police."

The man hesitated.

"What will it be?" the bartender asked.

Liko wondered if this was the same bartender from the previous night, or a different one. This guy was doing his job.

Jack lowered the peineta and the man backed up, two quick steps.

The assailant spit on the floor in the direction of the bartender, Jack, and Charla, then turned and walked away with the other man and woman following him. They walked into the room where Liko was standing and they took a seat at the table next to his. They sat down without beers.

Liko was surprised that they had retreated all the way to his quiet, dark part of the bar.

"Fuck," said the man who was soaked, "we have no beer."

"That's not my fault?" the second man said.

"Get us some beer," the man ordered the woman.

"Oh, yeah," the woman said. Liko could hear the sarcasm in her voice. "You think he's going to serve us after you spit at him?"

The man who was soaked with beer turned in his chair and faced Liko. "What are you staring at?"

"Your pants are wet," Liko said. "What did you do to yourself?"

The man jumped up from his wooden bench and started across the floor towards Liko. The other man and woman grabbed him and pulled him back. Together they directed him towards the front door. "Let's go," the man said, urging him on. "I've got some beer at home. And some weed."

The assailant cussed at Liko. The three then turned and walked to the exit.

As the door closed behind them, Liko turned his attention back to Charla and Jack. They had moved from the bar and had taken a booth together. *Nice*, Liko thought. *Jack has to know that Charla is trans, so she should be OK.* Liko had no doubt that Charla could handle Jack by herself.

A waitress passed by and noticed the spilled beer and broken glass on the floor. "I'll take care of that, hon."

"Thanks." After paying his bill, Liko got up and left the bar and walked back to his condominium.

21

Taxi Ride

That night Liko dreamed Marimichael jumped into the back seat of his taxi just before it pulled away from the entrance to his condominium. Liko was flabbergasted. *She must have nerves of steel*, he thought. She entered from the street side, coming from the Ritz Carlton.

"Hi," she said, sliding into the back seat, placing her hand atop his.

"What the hell—"

She smiled and patted his hand.

Her smile disarmed him.

He turned to the taxi driver in the front seat. "To the movie theater. The one in Georgetown. Somewhere off M Street." That wasn't his original destination, but the shock of Marimichael's sudden appearance had totally discombobulated him and he couldn't remember where he was going; the movie theater seemed as good a destination as any.

He turned sideways in the seat to squarely face her. "What are you doing here?"

"Join us?"

He lowered his voice. "Are you kidding?"

He motioned with his hands for her to lower her voice, too. In a whisper he asked, "Are you involved in what's happening?"

"The rebellion? Yes."

"And the killings?" He hoped she would say no.

"Non-violent, civil disobedience does not work against the fossil fuel industry. They are too wealthy, too powerful, and too entrenched."

She smirked and then continued. "It's sad, though. Augustus could have done the right thing, but he cared more about his money than his wife."

"So now you use violence?"

"Force is the most effective way to stop a monster. Violence must be met with violence."

"I don't accept that."

"Democracy, even at its best, requires compromise and takes a long time. We no longer have the luxury of compromise and time. The climate crisis must be addressed now. Today."

"So you *are* a part of it?" he whispered. "The hacking of congressional websites and corporate databases? The kidnapping of Heidi Dimer? The recent assassinations?"

She dropped her voice to a whisper. "Yes."

"Including Derichenko?"

"Yes."

"Why?"

"He would have killed you, Liko. After what happened in Davos, it was just a matter of time."

Liko wondered if that was true.

She squeezed his hand. "We wouldn't have to do this if the fossil fuel industry hadn't lied to us for thirty years. They knew thirty years ago that they were hurting the planet. They knew what would happen, yet they lied. The fossil fuel companies lied to us.

"Liko, time has run out. People in every country around the world must organize, must mobilize, and, if necessary, must bring their governments down."

The taxi driver pulled behind a long line of cars in the far right lane heading down M Street into Georgetown.

"Join us," she repeated.

"I can't," he said. "I just can't."

For the next block she stared blankly out the window. When she turned again to face Liko her eyes were red-rimmed and teary.

"Stop!" She pointed at the next corner and exclaimed, "Here! Right here!"

As the taxi pulled to the curb, she hopped out and ran across the street. She caught the bus, the Georgetown Circulator, heading in the opposite direction.

Liko stared out the window of the taxi at the bus. The taxi switched back into the inside lane and rejoined the long line of cars.

Liko wished that he had gone with her.

"Let me out!" he exclaimed.

As the driver pulled curbside again, Liko dropped a twenty-dollar bill into the front seat and said, "Thanks!" as he jumped out of the taxi.

He looked back down M street, but the traffic was lighter on that side of the street and the bus was gone.

"What am I to do?" he said under his breath.

Liko woke up.

Change is coming whether we want it or not, he thought. *It is coming either by revolution or by climate disaster.*

22

Ambush

The next morning, Liko phoned Jack and asked him to meet for lunch at Café Deluxe, catty-corner to his condominium building. Jack said he was busy, but he agreed to meet, reluctantly. When he finally arrived, the waitress showed him to Liko's table.

Jack was very conservatively dressed in a brown business suit and white shirt—no tight pants, no purple tie, no peineta in his hair. His hair, though, was still perfectly coiffed.

Jack ordered a roast beef sandwich with fries, and he asked for coffee and a piece of pie. Liko ordered the restaurant's equivalent of a farmer's breakfast. The food arrived quickly.

"You don't look so good," Jack said.

"I saw Marimichael last night."

"Damn! Why didn't you call me?"

"It was only a dream." He saw the disappointment in

Jack's face. "She jumped in my taxi and then jumped out. And then she disappeared."

"That's crazy."

"She wanted me to join her."

Jack stopped chewing and looked up at Liko. "And?"

"I said I couldn't." Liko pushed the toast through the yolk on his plate. "It goes against what I believe. I still believe in democracy."

"I see," Jack said.

Liko dabbed at the runny yolk. "I have something else to tell you."

"Another surprise?"

"Yes."

Jack picked up his coffee and held it in both hands. "I'm ready. Fire away."

"I saw you and Charla last night. At the Sports Grill and Bar."

Jack jostled his coffee. He picked up his napkin and wiped hot coffee off the back of his hand. "I saw you too."

A bit of yellow yolk missed Liko's mouth, falling from his fork onto his lower lip, where it flowed to his chin and collected. Liko quickly wiped his chin with his hand and then wiped his hand on his napkin. "Charla is a good friend of mine."

"You are very lucky."

"I know." Liko said.

He then decided to change the subject. "It's too bad the blackmail failed."

"Really?" Jack said.

"I wanted the Senate to vote on the Carbon Pricing Bill." Liko sopped up the last bit of yolk onto a piece of toast and popped the slimy mess into his mouth. "I wanted the bill to pass."

"But you didn't join the O'Briens' rebellion."

"No, I didn't. But we still need a revolution." He looked carefully at Jack, and then clarified what he believed: "Just not this one."

"Revolutions are... messy."

"But we have to do *something*. We need action. And yesterday."

"I still believe in non-violent civil disobedience. How about you?" Jack asked with narrowed eyes.

"I *want* to believe but how can I? All I see is gridlock. Chaos."

"Congress shall make no law ... abridging the freedom of speech, or of the press; or the right of the people peaceably to assemble, and to petition the Government for a redress of grievances."

Liko smiled. He had been warned that Jack liked to quote the Constitution. "So we should petition the government for a redress of grievances?"

"You don't look convinced," Jack said. "There is no justice in assassinations. Assassinating the fools who lied to us, or the monsters who continue to profit from fossil fuels, that will not roll back climate change."

Liko looked across the table at Jack. "Unfortunately, no."

"Anarchists—mostly men—have taken to the street," Jack said. "They're drinking and brawling. Anarchy will not roll back climate change, either."

"But without a revolution, what hope do we have?"

"Liko, revolution destroys the bad with the good, and nobody's smart enough to avoid all the unintended consequences of radical change." Jack added, "You don't look convinced."

"I'm not."

"Revolution risks destroying everything and turning over all our resources to the mob and to anarchy. It is very dangerous. The people who advocate it are irresponsible."

"Irresponsible?" Liko asked.

"Yes," Jack repeated. "Irresponsible. An example is the French Revolution. It started out great and then was taken over by the mob, and Robespierre. Another example is the communist revolution in Russia. It was started by a bunch of intellectuals who were motivated by wanting... by hoping...to relieve poverty and to give greater justice to everyone. And then those people were all killed and kicked out when the radicals like Stalin took over. *They were killed.* That's what the Trotsky business was about. And most revolutions have ended in that kind of loss of control."

"What about the American Revolution?" Liko asked.

"The American Revolution was an exception."

"Why was it an exception?" Liko asked. "Why didn't the American Revolution descend into anarchy like the French Revolution?"

"I'm not exactly sure why it succeeded, but it did. And we never descended to pure democracy. The thinkers maintained some control—representative democracy. And so, I think possibly it was just dumb luck. But it was a good thing. And maybe we are losing it all now in this terrible change in our democracy."

When Liko didn't ask another question, Jack added: "People who want to overthrow the system, who want to change people's way of life suddenly, they don't know what they are doing. They are risking a lot, and we shouldn't allow it. We should resist it. People are scared."

Jack waited patiently, eating his pie, while Liko finished his farmer's breakfast. Jack insisted on paying the bill, and as they walked to the front of the restaurant, Jack asked, "Did you ever see the movie *The Day the Earth Stood Still?*"

Liko nodded yes and said, "Klaatu barada nikto—or something like that."

Jack laughed. "Remember when the alien shut off all the electricity everywhere on earth?"

"Yes," Liko said. "It was very dramatic. And effective. He got everyone's attention. The cars, the elevators, everything stopped working."

"Well," Jack said. "Maybe this rebellion has gotten everyone's attention." He slapped his hands together, flamenco style, the first three fingers of one hand striking the palm of the other, making a sharp and snappy sound. Liko flinched. "Perhaps it will scare Americans out of their complacency."

"I hope so," Liko said.

As they stepped out of the restaurant and onto the sidewalk, Jack commented: "It's a cloudy, dreary day." He pulled his wool coat around himself.

He stopped suddenly and threw his arms around Liko and gave him a heartfelt hug. "I know why you asked me to lunch."

"Why?" Liko asked. He wasn't sure himself; nevertheless, he had felt the need for a meeting.

"Because you are concerned about Charla." Jack patted Liko on the back, then pushed him away and held him at arm's length to take the measure of him. Jack smiled. He was a small man embracing a Hawaiian twice his size. "Because you want to know my intentions. Am I right?"

The bullet ripped through Jack's back and exited through his chest. His body slumped onto his knees in front of Liko.

Spattered with blood, Liko leaned forward and shoved

his hands under Jack's arms and pulled him back into the entryway. He twisted Jack onto the ground and then dropped down on top of him, covering him.

Another shot struck nearby. One passed through the top edge of Liko's goose down coat. Liko watched several down feathers floating in the entryway.

And then it was quiet.

23

Hospital Visit

When Liko and Charla visited Jack in the hospital, it felt like attending a wake. Jack wasn't dead, but he was in a coma. His friends and colleagues brought gifts of food and knickknacks instead of flowers, because for some unknown reason, word had circulated that he had an allergy to flowers.

Visitors dropped off baskets of fruit, boxes of chocolates, and even homemade cookies. The gifts of food spread out on the spare bed in Jack's room. Gifts of wine and whiskey rested on shelves in his clothes closet, out of sight.

Charla bought a stuffed tiger in the hospital gift shop and set it next to Jack's bed. She hoped that the bright orange and black, striped plush toy, would cheer him.

Jack had no children. His closest relatives lived out of state and were unable to visit. Consequently, Charla took it upon herself to sit beside Jack's bed, waiting for him to wake up. She greeted one friend after another who stopped by to pay their respects to a man they admired, but who they expected to soon die. Liko, though, remained hopeful.

As folks visited and talked story, Charla kept vigil. When Liko visited, he sampled the food, especially the homemade perishables, and he sipped the whisky. He didn't think Jack would mind.

One visitor after another told stories about Jack. One visitor said: "Jack's a true patriot."

Some guests talked in the past tense, which began to unnerve Liko. One said: "He was only happy when America was living up to its ideals." Another said: "He was loved by everyone who knew him."

Franz, the FBI man who had been with Liko and Jack in the Black Hawk helicopter and at the coal plant, brought a deck of cards. "For when he wakes up," Franz said. "Nothing worse than sitting in a hospital bed bored, trying to find something on TV. Huh. Huh-huh."

Franz added, "Jack's a great poker player, so watch out! He can pull cards from the air or make them vanish. Huh-huh-huh."

Liko thanked Franz for the warning.

Franz then told Liko, "We'll get her" — he meant Marimichael — "and when we do, I'll give you a call."

Jack lay unresponsive, lost, deep in his coma. Charla wanted to be present when he woke up, so she continued to sit beside his bed from early morning to late at night, until the nurses ordered her to leave.

Three days passed. Jack remained in a coma.

In the meantime, Liko witnessed the escalation of violence in the District. Environmentalists blocked the roads leading into the city. Traffic jams caused delays in emergency services. Ambulances could barely move.

Protestors pushed abandoned cars, boats, and heavy appliances like refrigerators and deep freezers onto the streets. Bulldozers were dispatched to remove debris that blocked the streets. The dozers pushed abandoned boats and appliances to the side of the road, scraping the hell out of them. Vehicles were towed.

One day anarchists randomly destroyed transformers, taking down sections of the city's electric grid, partially shutting down local and state government. Businesses closed. Emergency generators at federal buildings turned on.

It had been a long, hot summer in DC and so far, a dry winter. Trees and plantings burned like kindling. Arsonists set cars on fire. Businesses burned to the ground. Liko saw one man brandishing a sign that read "LET IT ALL BURN!"

Protesters and riot police clashed. Tear gas choked protestors and water cannons knocked demonstrators off their feet.

Every evening there was a curfew. It didn't affect Liko because he left for home at 8pm, when visiting hours were over, and it was only a three-block walk from the hospital to his condominium.

Each day, those who were drawn to chaos celebrated in the streets. Supermarkets were looted. Storefronts were smashed. Christian evangelists stood on street corners and preached "The end times are here at last!" and "Praise the Lord!" Others chanted "Tear it down!"

All public transportation shut down. Even the Metro trains stopped running. Flights out of Dulles International Airport and Ronald Reagan Washington National Airport were cancelled because pilots and crew could not make it to work.

The National Guard patrolled the streets, and the army, out of Joint Base Myer–Henderson Hall, mobilized to the White House, the Capitol, and the Supreme Court Building. The conflict had escalated to a national emergency: President Clincher signed a Proclamation on Declaring a National Emergency by Reason of Certain Domestic Terrorist Acts.

One evening, on his way back to his condominium from the hospital, Liko stopped a woman from attacking the equestrian statue of George Washington in the center of Washington Circle. The statue was high up on a concrete pedestal. The woman couldn't reach General Washington, so she was striking the feet of his horse instead, again and again.

Liko grabbed the crowbar out of her angry hands. He ordered her to leave and to go home. She yelled obscenities at him and called him a thief when he refused to return the crowbar.

People are going crazy! All we need now is for a war meteor to streak across the sky and explode into fireballs.

When Liko entered his condominium he was met by Charla. It was around 9 o'clock in the evening.

"Thank god you're home," she said.

Liko collapsed into his Italia chair. "What happened?"

"It's crazy out there," she said.

"I know." Liko couldn't believe what was happening in the District, either. He told Charla about the crazy woman attacking George Washington's horse.

As they talked, Liko stared across the room at the abacus lamp stand. He recalled how much Jack had admired it. *When he gets out of the coma, maybe I'll give it to him. He'd like that, I bet.*

He recalled Franz's comment about Jack playing poker: "He's not a card shark, but he knows some tricks. He can pull cards from the air or make them vanish. That kind of thing."

Liko sat up straight in his chair, startling Charla. "I think we've been—"

"Been what?"

"Sssh!" Liko placed his index finger to his lips. He pushed himself up from his chair and walked over to the abacus lamp. He moved the sliding beads back and forth until he discovered a small electronic device—a microphone!

The FBI has been listening to our conversations.

Charla was furious. "Just wait 'til he wakes up!"

24

Oil Refinery

The eco-terrorist sat on the inverted five-gallon plastic bucket reading a graphic novel, *Wild Blue Yonder*. His Apple watch buzzed: 2:30am. He stood up, closed the novel, and set it on top of the bucket. He stretched the muscles in his back, bent and twisted at the waist, and finished with three quick squats. He rotated his head side to side.

He walked across the room to the racks of servers and hubs and computer cables and unplugged a bank of servers and then cut the electrical cords with wire cutters so they couldn't be plugged back in. He proceeded to the next rack of phone cables. He focused on a box labeled "external phones," tracing the phone cables back to a common junction where they all came together before disappearing into a circular opening in the wall. He grabbed the mass of cables in one hand and cut through them with the wire cutters and then slipped the wire cutters into his back pocket.

He walked past more racks of servers to an exit door and stepped out into a hallway. Seeing no one, he quietly closed the door behind him. The sign on the door read "Electrical Room. Authorized Personnel Only."

He walked down the wide hallway until he came to a bank of interior windows that looked into a control room. One man in the room was looking at a console and the other man looked up at the terrorist. An inquisitive expression filled the worker's face.

The terrorist proceeded down the hallway, opened the door to the control room, and entered. He pulled a gun from the shoulder holster that was hidden under his tan workman's jacket. "On the floor!"

Both workers scrambled from their chairs to the floor.

"On your stomachs!"

The workers rolled onto their stomachs.

"Hands behind your back!"

The terrorist stepped to one and then the other control room operator, slipping long, white plastic ties over their hands and cinching them tight.

An alarm sounded and a pre-recorded voice said over the refinery's speaker system: "All personnel evacuate immediately. This is not a drill." A red light attached to the wall began flashing.

Out of his left back pocket the terrorist took a small cellular phone stand, which he set on a counter. He attached his cell phone to the tripod and stepped behind the phone, focusing the picture so that the two men on the floor, the consoles, and half the control room were all in the picture.

Standing behind the phone and out of the picture, the

terrorist pushed a small remote-control button and his disposable phone took a photo. Using the remote, he reset the phone to video mode and pushed record. He silently let the video run for four seconds and then, in a loud voice, he yelled at the men on the floor, "Don't move or I will shoot!" As they lay without moving, he let the video run for another eight seconds before stopping it.

The terrorist removed the phone from the small tripod. He uploaded the photos and video to a website. He then discarded the phone onto the counter. He pulled the wire cutter out of his pocket and tossed it onto the counter, too.

He pulled up his shirt, revealing a broad money belt wrapped around his waist holding explosive charges. He undid the money belt and draped it across two of the console units. He then pushed several buttons on a device attached to the belt and a timer started. He took several steps backwards.

"On your feet," he ordered the workers. "Stand up!"

While pointing the handgun at the men, he smiled and said, "This gas refinery is going to explode. You should run like hell." He looked at them and offered some additional advice: "Run away from the river. I have placed explosives there, too." He motioned them through the door.

"Run!"

They took off running down the hallway, their hands still cinched tightly.

The terrorist pulled the control room door behind him and, after making sure that the door lock was jammed so it could not be reopened, walked quickly down the hallway in the opposite direction of the workers. He went directly to a side entrance of the building where two other eco-terrorists were seated in an electric cart, waiting for him.

He jumped onto the empty bed of the cart and all three men sped away, as fast as the electric cart would go, towards the river.

When the three eco-terrorists reached the river it was 2:45am. The men hopped from the cart and ran to a chain link fence and slipped through a hole they had cut earlier.

A fourth man sat waiting for them near a rubber raft. Together, all four men pushed the Zodiac into the water and climbed aboard. They then motored to the middle of the river where they did an abrupt U-turn to face the refinery.

One man sat with his hand on the marine motor and kept watch. Each of the other three men opened a box and removed a drone, a sparkler, and a lighter.

Meanwhile, several hundred miles away, a computer savvy eco-terrorist downloaded the photo and video of the control center from the website. "I'll have your background ready in a few minutes," he told Marimichael.

She was seated in front of a green screen background. A directional microphone sat in the middle of a table in front of her. The Rode VideoMic was wrapped in a furry windshield to eliminate extraneous sounds.

The computer technician married the image of the control room with the image of the green screen, effortlessly. After adjusting the quality of the new image, he told Marimichael, "We're good to go. You're sitting in the control room."

"I'm ready." She straightened her back, smiled, and looked directly at the microphone and another disposable phone.

"Running the first video. On my count," the technician said. "Four, three, two, one."

Marimichael appeared on screen as if she were occupying the control room. The oil company's logo, control panels, and monitors appeared behind her. The two control room operators lay face down on the ground in front of her, their hands cinched behind their backs. The pre-recorded voice, off-camera, ordered, "Don't move or I'll shoot you!" The refinery alarm and warning message repeated over and over: "All personnel evacuate immediately. This is not a drill." The red warning light flashed an ominous red glow onto everything in the control room.

Hundreds of miles away, the technician waved his hand downward while pointing at Marimichael. He said in calm voice:, "Action!"

"Who are we?" Marimichael asked, looking into the camera phone. "Our group has no name. In fact, many have called us the No Name Climate Crisis Group. We like that. So, from this day forward, we are the No Name Climate Crisis Group." Marimichael smiled.

Her smile vanished and her face filled with determination. "We are a movement to stop the climate crisis from destroying the natural world as we know it.

"We are veterans and warriors who are fighting for our lives, the lives of our children, and the lives of our children's children. We are fighting not only for humanity, but also for the vanishing ark—all the wonderful species of animals and plants that are being driven into extinction.

"You have a role to play. Register to vote. Elect leaders who will fight to protect our future. Join us!"

She pushed her chair back from the desk and stood up. "And now we are going to destroy this monstrous refinery."

In the high-octane gas production unit of the 150,000 barrels/day gasoline refinery, the hydraulic lift turned on. The mechanism was simple: a car battery, a timer, and an electric lift. The had placed the hydraulic lift beneath a pipe elbow that carried propane gas.

When three thousand pounds of force pushed against the pipe elbow, it ruptured and liquid propane escaped, hissing, vaporizing.

The propane, heavier than air, settled onto the concrete floor, turning white as moisture condensed from the air. The propane plume spread, rolling along the floor like an inflating donut. The thick, dense plume advanced in rolling white waves until it blanketed the entire floor of the high-octane production unit.

An alarm and intercom system had been blaring, warning the night shift to evacuate. The crew fled, as fast as they could, abandoning the facility. The refinery's security system automatically notified the police and fire department. Emergency medical services began their mad rush to the refinery.

The terrorists released the drones, one after another. The drones flew along a pre-programmed route, using GPS coordinates, to the high-octane unit. Each carried a white sparkler burning at 3,000 degrees Fahrenheit. They burned bright, throwing off sparks as they flew into the propane plume.

The first explosion destroyed the high-octane gas production unit. Crows, roosting along the property line in a row of trees, burst skyward, spooked. Confused, they circled in the moonlight, making corkscrew patterns at the perimeter of the refinery.

The second explosion was massive, as hydrocarbons from a processing tank erupted like a volcano. Rolling black smoke engulfed the crows. Yellow and orange flames singed their wings.

The men in the Zodiac felt the shockwave of the refinery explosion on their faces. One grabbed his baseball hat, which was his souvenir from the Nationals victory a few years past.

The shock wave tore through the tree line, bending and snapping trees and rattling houses as far away as Kentucky. Dogs hid beneath their owners' beds. Knickknacks fell off shelves. Horses in barns awoke, their ears flicking and upper lips curling. They breathed heavily, examining the air for danger. More than a hundred miles away, the shock wave rattled a metropolitan museum. Priceless porcelain pieces fell off stands in display cases and broke.

A piece of the hydrocarbon processing tank, the weight of four Ford F-150s, flew a distance of eight football fields and landed on the other side of the Firebrand River. The men in the Zodiac ducked, reflexively, as the projectile flew overhead. Smaller debris landed everywhere within the 850-acre refinery.

The fire burned for thirty-six hours until it self-extinguished. Video of water hoses aimed at the base of the fire looked like frogs pissing into a large campfire.

"Today we have crippled a major gasoline refinery," Marimichael said. "Make no mistake: from this day forward, violence against the earth will be met with violence against those responsible. The fossil fuel industry is a house of cards," she said. "Join us in the climate crisis revolution."

Gasoline futures jumped before the end of the day.

25

Motivations

Jack remained in a coma and Liko felt miserable. To console himself, he decided to call his great-aunt, who lived in Black Point Beach. She had always been there to help him, and he needed emotional support, desperately.

"How's the weather in Hawaii?" Liko asked.

"Same as you remember it," she answered. "How's the weather in DC?"

"Terrible," he said. "It's either too hot or too cold, or too dry or too rainy."

"Have you made friends?"

"A few. Danielle, who used to write for the *Sun*. She now teaches journalism classes and history at West Virginia University, and she has her own blog. And I've gotten to know the concierge and the chief mechanical engineer in my building. And Charla, a friend from Maritauqua Island, is visiting."

"I'm glad to hear that you are meeting people. Do you like your apartment?"

"I do," he said. "Especially the location. It's close to everything: the metro, a movie theater, and several museums, including the National Geographic and an art gallery called the Phillips. My condo is within walking distance to the Kennedy Center, too."

"Are you eating OK?"

"I'm not suffering. There are restaurants in Georgetown and Dupont Circle, just a short walk from my condo. Trader Joe's and Whole Foods are nearby, too. It would be nice, though, if we weren't under martial law and a curfew."

"Are you safe?"

He wasn't sure how to answer that. She didn't know about his visit with the eco-terrorists, except for the news event at the coal plant that had been carried nationwide by CNN. Nor did she know that he'd been riding around West Virginia with Danielle, ferreting out information about the terrorists. And he hadn't told her about Charla's boyfriend, Jack.

Should he tell her that Jack had been shot while standing next to him? He didn't want her to worry. What good would that do?

Liko took a deep breath and spent the next ten minutes telling her everything that had happened. He'd always had that kind of a relationship with her.

He was glad she didn't ask him if he had a girlfriend, but before he knew what he was saying, he told his her everything about Marimichael. He even described some of her tattoos, including the meteor that blazed across her back.

She listened, and then she asked a lot of questions. He answered all of them the best he could.

"I hope you can visit soon," she said.

"I'd like that," he replied. "I miss you."

"I love you, Liko."

"I love you too, Auntie."

"Goodbye for now."

"Goodbye."

Charla had been watching television while Liko talked to his great-aunt, and when he hung up she asked, "What does the tattoo look like? The meteor?"

He described it and said, "The tattoo artist was inspired by Frederic Church."

"Who was the tattoo artist?"

"I don't know. But you can Google a picture of the painting."

Charla nodded. "Why would she get something like that? A meteor tattoo across her shoulders?"

"There is a connection between the Great Meteor and John Brown. The meteor symbolizes John Brown's martyrdom and the start of the American Civil War. He was an abolitionist who fought to free the slaves."

"Where could you get a tattoo like that?"

"I have no idea," he answered. "Why, do you want one?" He knew that she didn't. He was just teasing her.

Suddenly Liko leaned over and whispered softly in her ear, so the microphone would not pick up his voice: "I think Danielle would be interested. Maybe interview the tattooist?"

"Maybe the artist has a picture of Marimichael's tattoo?" Charla whispered.

"That would be great!" Liko said, raising his voice.

Liko met Danielle at Dumbarton Oaks Museum and Gardens. He hoped it would be a safe and quiet meeting place, hidden away from all the current craziness in the city. They discovered, unfortunately, that the museum and gardens were closed, so they went to the nearby Dumbarton Oaks Park. They accessed the park from R Street and entered along Lovers Lane.

"Did Marimichael say anything about her tattoo artist?" Liko asked. "His name? The name of his shop?"

"No. Why?"

"I think if we find the tattoo artist, you find a good interview." He grinned. "And maybe the tattoo artist keeps pictures of his work?"

Danielle's eyes widened.

"I've seen her tattoos, and they're beautiful. If he has pictures—"

26

Tattoo Shop

Liko and Danielle visited every tattoo shop in West Virginia looking for an artist who had custom designed a meteor. They spent a week on the road, visiting one shop after another. They had no luck.

They were now in Morgantown and had just visited the last of three shops.

Exhausted and hungry, they parked in front of a mom and pop restaurant on High Street. The words Holly Cafe and a sprig of holly were painted on the window glass. Liko fed three quarters into the street meter and they went inside. It was 2 o'clock in the afternoon.

Liko surveyed the café. A waitress worked behind a short counter, pouring coffee, serving customers seated on swivel stools. A row of booths with wide seats and high backs, upholstered in shiny brown Naugahyde, ran from the counter to the back of the restaurant and the restroom signs. Along the opposite wall and in the center of the café, several row of tables and chairs faced the booths, arranged to accommodate single customers or small families. A red carnation in a small vase smiled from the center of each table. The middle-aged waitress came from around the counter, grabbed two menus, and directed Liko and Danielle to follow.

Danielle quietly elbowed Liko. She pointed to a colored pencil sketch of a meteor framed on the wall behind the front register — in plain sight where everyone entering the restaurant would see it.

Liko almost tripped. He caught hold of Danielle's shoulder to steady himself.

The waitress directed them to an empty booth. It was clean, comfortable, and cozy. The seat was well worn, and so was the menu, which offered down home cooking: chicken fried steak in cream gravy with mashed potatoes and roasted green beans, chili with cornbread, grilled cheese, BLTs, a variety of burgers and fries, and a golden delicious apple pie.

After ordering their meal, Danielle asked the waitress, "I like the drawing by the register, the fireball meteor. Did the owner draw it?"

"Oh, my son did that," she said. "Everyone comments on it."

"He did a nice job. It looks like the original."

"The original? Do you mean his tattoo design or the Frederic Church painting?"

Liko spit out his coffee. "Your son does tattoos?"

"He's an award-winning artist." She flipped the page of the order form.

"Does he have a shop here?" Liko asked.

"No," she answered. "He moved to DC. Took his tools with him. Opened a parlor downtown."

"Whoa! How long ago?"

She looked at him suspiciously.

"I'm sorry," Liko said. "I find a meteor tattoo fascinating. Does he still do them?"

"The Great Meteor? No, it was a custom design. He did only one."

"A special client?"

"A college student at WVU."

Liko sighed, suddenly realizing that he should have checked out the tattoo parlors near Marimichael's alma maters first, instead of driving all around the state.

"Do you know where I could find her, that client?" He added, "I'd like to see the tattoo, if she would let me."

"She's long gone, honey." The waitress laughed. "She graduated and moved away."

"How can I reach your son?"

Again the waitress looked at him suspiciously. She tilted her head and balked at his request. "He's hard to reach."

"If I give you my name and number, can you ask him to call me?"

"Well..." she pressed the point of her pencil into the pad, "I guess there's no harm in that." She handed him her order pad and pencil. "Write down your contact information."

Liko wrote as quickly as he could and then handed the pad and pencil back to her. "Thank you."

After eating their excellent burgers, Liko paid at the front register. As they approached the front door to leave, the restaurant business license caught his attention, next to the food inspection rating. The name on the license and the signature on the inspection report were the same: Kirby Long.

Liko walked back to the register and stepped behind the counter to take a closer look at the meteor drawing. The artist had signed Cody Long.

"What are you doing?" Cody's mother asked Liko.

"Admiring your son's work," Liko said. "The realism. The color. The design." Liko meant what he said.

"You can't be back here," a gruff voice said.

Liko turned to see a man with a food-stained white apron wrapped around his waist. He had a deep, commanding voice as if he might be the owner of the café, which would make him Cody's father. He was handsome, with coal black hair, oiled and carefully combed. His frame was lean and wiry, and he had almost no hips to hold up his pants, but his muscular forearms revealed he was strong and not to be messed with. His eyes were as black as his hair, yet kind.

"No problem," Liko told him. "I'm leaving." He stepped out from behind the counter. "I admire your son's work, Mr. Long."

To the waitress Liko said, "I hope to talk to your son. Please give him my number. Thank you."

And with that, Danielle and Liko left the café.

"You have the makings of an investigative reporter," Danielle said.

"I did OK?"

"Not bad, Liko. Not bad at all."

On the drive back to DC they were quiet and listened to the radio most of the trip. Their first stop in the district was North Capitol and I street, the DC Department of Health's Vital Records office. Danielle requested and received a list of the dozen or so piercing and tattoo parlors operating in DC.

She ran her finger down the list until she came to Cody Long. "The name of his shop is Rebellion Parlor DC." Danielle and Liko looked at each other for a long moment, struck by the word "rebellion" in his shop name. Maybe it only referred to the rebellious reputation that tattoos can have.

Liko Googled the parlor. The shop website had a biography of the tattoo artist. "Want to see a picture of Cody?"

Standing beside Liko, Danielle leaned in so she could see the iPhone screen. "He has a birthmark on his left cheek."

Liko looked closer at the photo. The birthmark extended to the corner of Cody's upper lip. He had seen a birthmark like that on a young boy in the thumb drive materials that Marimichael had given him. Liko read highlights from the website: "He does custom, one-of-a-kind designs. No body piercing. Appointments only. No walk-ins. His work can be viewed on a linked Instagram account. He has T-shirts and art prints for sale."

"The shop closes at 6pm." Danielle checked her watch. "It's already 6:15. We'll visit tomorrow?"

"That works for me." Liko was tired and looking forward to a long, hot shower when he got back to his condo.

"What time does he open?"

"Noon."

Liko drove Danielle home. He then drove to his condominium and parked in his assigned parking space in the underground garage. After his hot shower, he felt restless. Charla was not home and he felt bored. He turned on CNN.

He glanced at his watch: 8pm. He thought about the compliment Danielle had given him: "You have the makings of an investigative reporter." He felt proud of himself.

Why not surprise Danielle, he thought. *I can scope out the parlor, take a few pictures, and get back home before the curfew. Maybe I can get some good nighttime shots.*

He donned his winter coat, boots and skull cap. As he stepped out the front door of his building, he pulled on his leather gloves. He didn't like walking with his hands in his pockets. He clinched his hands against the cold, stretching out the glove leather.

Liko hopped on the Foggy Bottom Metro, transferred to the red line at Metro Center, and jumped off at the Gallery Place Metro, next to the National Portrait Gallery. The tattoo parlor was a few blocks north and several blocks east, an easy walk.

27

Rebellion Parlor, DC

The parlor door was locked, and no one answered when Liko knocked. The building was typical brick, uninteresting, and without signage, just the numbers for the street address. *A waste of time*, he thought.

He looked up the phone number for the tattoo shop from the list crumpled in his pants pocket. He expected to get a recording, and he intended to make an appointment for tomorrow. If he could arrange things for the next day, perhaps he would impress Danielle.

A man answered on the second ring: "Rebellion Parlor. Can I help you?"

Surprised that a human voice answered, Liko said, "I'd like to make an appointment."

"Our first opening is this Wednesday."

"I need something sooner, something tomorrow."

"Sorry," the voice said, "but that's the best I can do."

"Is this Cody Long?"

A pause and no answer.

"My name is Liko Koholua. I was in the Holly Café today and saw your drawing of the Great Meteor. I asked your mother to call you and to give you a message." Liko slowed down, so Cody could process what he was saying. "Did she call?"

"Yes," the voice answered. After an extended pause the voice added, "She did."

"So I'm talking to Cody?

"Yes. What can I do for you?"

"I'm doing a story on Marimichael and would like to talk to you about her tattoos."

"What did you say your name was, again?"

"Liko."

Once again there was a pause.

"Actually, I'm standing right outside your parlor door."

"I don't do walk-ins."

"I understand. I apologize. But can we meet tomorrow?"

"Let me call you back in a few minutes."

"OK, I'll wait for your call."

Liko paced up and down the sidewalk, hoping that he hadn't screwed things up. He saw nothing green, nothing alive, except one rat and then two, scampering along the foundation of the next building up the street, now stopping, now darting, now standing on sturdy hind legs, front feet dangling like a *T. rex*, large ears alert, listening, and then dropping to all fours before disappearing into the side of a cracked concrete stairway. Liko shivered.

After several minutes, the front door of the parlor opened and a man stepped out.

Liko was startled; he thought the business was closed

for the night so he was surprised to see a man suddenly appear in the doorway. A man who pulled a gun from a shoulder holster.

Liko threw up his hands. "I'm unarmed."

The man motioned Liko inside. Liko obeyed.

The man then motioned Liko up the flight of stairs to another door. When they reached the top of the stairs, the man rapped hard.

The door opened from the inside. Another man appeared in the doorway and gestured for Liko to enter. He had a birthmark on his left cheek.

Yes! Liko thought. "Cody?"

The man said nothing. The man with the gun, though, pushed Liko from behind into the shop's waiting room, and then closed the door behind them.

"Thank you for seeing me," Liko said.

"Have a seat," the man with the birthmark said, motioning Liko into one of three chairs. Copies of *Tattoo Energy* and *Tattoo Life* lay on a side table.

Liko sat.

"Are you by yourself?" the man asked.

"I am," Liko said. "Are you the artist who did the meteor?"

"How do we know that you're alone?"

"You don't," Liko said. "I've been working with a reporter, but she doesn't know I'm here."

The man with the gun pointed it at Liko's chest—a target he couldn't miss at such close range.

"What do you want?"

"Cody, I met your parents. Kirby makes a great hamburger and fries. Your mother is proud of your work."

"You met my parents?'

"Yes, earlier today," Liko said. "At their Holly Cafe."

Cody's cell phone dinged. He read the message and then directed Liko: "Please come with me."

Liko followed Cody across the waiting room and through a door into a short hallway. They passed a room on their left set up for tattooing and entered the next room.

A video camera was set up on a tripod pointing at a green screen. To the right of the camera, and out of view, was an autoclave for sterilizing instruments. On the counter were boxes of single-use needles, sterilized gauze, and implements that Liko guessed were used in the tattoo craft.

Marimichael entered through a side door. He glimpsed an inflated mattress behind her, on the floor, covered with unmade sheets and a blanket. She closed the door behind her.

"Damn!" Liko said. He backed up a step and stared at her. She was wearing a short top, tied behind her neck and lower back like a hospital gown, and open in the back. The elastic band of her black EMT pants hugged her hips. The military-style boots she wore gave her an extra inch of height, so they looked at each other eye to eye.

"What a surprise!" She took a step towards him, both arms open wide, and gave him a heartfelt, genuine hug.

Liko's body tensed and stiffened. He glanced at Cody and the man holding the gun. They were surprised, too.

Marimichael held him for a moment. There was no softness, no gentleness in her hug. It was firm and uncompromising.

His hand came up and his fingers touched the bare skin in the middle of her back.

She flinched and pushed him away. "My new tattoo." She quickly changed the subject: "You weren't followed?"

"I don't think so, but I'm not good at these things." He tried to smile, but he failed. "I didn't know you were here, so they can't, either."

Marimichael turned to Cody and the man with the gun. "Put that away," she told the man. "Cody, we need to do the photo shoot now, and then we need to leave for another safe place. Are you ready?"

"I can make it happen," he said.

"Good."

She dismissed Cody and the gunman, and Liko found himself alone with her.

As an icebreaker, he waved his arms around the room and ventured: "Is this where you faked the control room video?"

"Yes," she said. "It is."

"Well, I'm glad you were here and not at the refinery." Liko smiled.

She gave him a perplexed look.

"The explosion and fire would have killed you."

She laughed. "Are they still looking for my ashes?"

"Probably."

She looked him up and down, head to toe, toe to head. "You look good."

"You too," he said, and he meant it. Liko hesitated, then said, "Did you try to kill me?"

"Liko! How could you think such a thing?"

Liko thought about Jack, who was still in the hospital in a coma.

She smiled and added, "If someone tried to kill you, I can say for a fact that it was not Derichenko." Her smile broadened. "He's dead."

Liko looked into her eyes and he knew that she was telling the truth.

"You killed him?"

"For you."

He sighed heavily.

"It doesn't make you happy?"

"No!"

"Are you here to join us?"

"No. I want nothing to do with all the violence that surrounds you."

"Then why are you here?"

He looked into her eyes. "Aren't you tired of the violence? The killing?"

A knock came at the door. "Come in," she said.

It was Cody, holding a Nikon camera in his hand. "We're ready if you are?"

"I am," she said.

They took the stairs to the first floor, exited the front of the building, and walked a block to an adjacent hotel, passing in front of the stairs where Liko had seen the rats. Cody led the way, pulling a large, black duffle bag on wheels. Marimichael, Liko, and the two men with the concealed handguns followed.

Marimichael had donned a light windbreaker. Liko had still not seen her new tattoo.

They entered the hotel lobby and rode the elevator to the Olympic-sized swimming pool on the third floor. There was only one hotel guest using the pool, and she quietly left when Cody politely asked her to leave. To prevent other guests from entering, the armed men taped "PHOTO SHOOT—DO NOT DISTURB" signs on the doors.

Cody unpacked the duffle bag and set up three cameras

on tripods. Liko stood against a wall, unsure what was happening and trying to stay out of the way.

Marimichael disappeared into the women's locker room. One of the armed men stood in front of the locker room entrance and the other took up a position at the main entrance to the pool, where they had entered.

Marimichael returned in a shiny white bathrobe. After a short conversation with Cody, she climbed a ladder to the top of the high dive, walked gracefully to the end of the board, and then turned her back to the cameras set up below.

"Ready?" Cody called from across the pool, focusing one of the cameras upward.

"I'm ready!" she called back.

"On my mark. Five, four, three, two, one," — he pushed a button and all three cameras simultaneously began shooting — "ACTION!"

She let Cody film the back of the robe for a full five seconds before turning to face the camera.

She let the silk robe drop from her shoulders. It glided down her nude body, over the edge of the diving board, and into the air beneath her feet. The robe fluttered to the pool, collecting on the surface as if it were a white lily pad.

Marimichael stood naked on the diving board. She turned her back to reveal her new tattoo: beneath the meteor an oil refinery exploded, hatched in black ink with white rolling clouds of gas and yellow-orange flames bursting skyward with the force of multiple explosions.

The meteor tattoo complimented the destruction of the oil refinery, perfectly.

Liko sucked in his breath. Marimichael and the artwork were breathtaking.

With her back facing the camera, Marimichael walked

to the edge of the diving board, bounced once, and was airborne: a graceful backward swan dive with a 180-degree turn mid-air.

A shot rang out. Marimichael's body twisted in the air.

Out the corner of his eye, Liko saw a man in a dark suit kneeling on one knee near the entrance to the men's locker room, a pistol aimed upwards at Marimichael.

Liko glanced back at Marimichael. Her body knifed into the pool water. She failed to pull up out of the dive. She went straight downwards to the curve of the pool. He saw her head strike the blue tile. He heard the water-muffled snap of her neck.

"FBI!" the man yelled as he rose to his feet and ran to the edge of the pool, his gun aimed to shoot Marimichael again. Liko raced around the corner of the pool towards him.

The FBI man pivoted, raising his gun to shoot Liko.

A shot rang out and the man was struck in the upper leg. The impact of the bullet knocked him off-center and he collapsed onto his knees as he pulled the trigger.

Liko felt the bullet pass in front of his eyes. "Fuck!" he yelled. Behind him, the bullet exploded the white ceramic wall tile.

Liko jumped onto the FBI man, knocking him to the ground, pinning his back against the floor. Liko slammed his left fist into the man's face, breaking his nose and jaw. Scared, angry, and surging on adrenalin, Liko raised his right hand. He focused on the right temple.

As he brought his fist down with all his strength, Liko saw fear flash across the agent's face. Liko shifted his upper body and shoulder, almost imperceptibly. His fist slid along the side of the agent's ear and head and slammed into the ceramic tile that was cemented to the concrete floor.

Liko heard the bones in his hand break as his fist smashed into concrete. The pain was so intense that he was unable to take a breath. He saw the look of surprise on the agent's face.

Something struck Liko between the shoulders.

When he opened his eyes, he had inhaled the smelling salts held beneath his nose. He jerked his head away and sat up, coughing.

"You were tasered." A paramedic towered over him. "Relax! Don't try to stand up or move."

Liko saw a syringe in the man's hand. Liko's sleeve had been pulled up, and he could now feel where the man had injected something into his shoulder muscle. It burned.

"I gave you a tranquilizer," the paramedic said. "If you try to stand or move around you will hurt yourself."

Liko felt a tingling in his right foot. He looked at his feet and discovered they were cuffed tightly with plastic cinches.

He started to complain, but then he saw Marimichael. The paramedics had stabilized her; she was on a stretcher with her neck in a brace and her eyes closed. Two long boards had been placed along her body, one on each side. Her legs, waist, and upper body were all strapped down to the stretcher.

He watched several men pick her up just high enough to slide her onto a gurney. A man covered her naked body with a light blanket, pulling it up to her chin. The gurney was raised up and she was wheeled out of the pool area.

One of the rebels who had escorted them lay outstretched on the floor at the entrance to the women's locker room. The other gunman lay on his back near the edge of the pool. Liko guessed that both men were dead.

Liko looked at the large gauze pads and ace bandages that had been wrapped around his hand. He could feel his hand pounding with each beat of his heart. He remembered smashing it against the floor. He now realized why his feet were cuffed and not his hands.

He did not see the FBI man he had punched in the face. Perhaps he had already been taken to an ambulance. Liko hoped that he had not hurt him too badly. Yes, he had shot Marimichael, but he was just doing his job and that's why Liko pulled his punch at the last second.

In the back of his mind, Liko realized that someone was missing from the scene: the tattoo artist, Cody Long. *Where was he? Did he get away?*

Liko turned to look at the cameras. The Nikon Z7 was missing from the tripod.

28

Capitol
Attacked

That evening, footage of the shooting of Marimichael appeared on multiple social media platforms. Facebook, Twitter, and even YouTube worked with the United States government to take down the footage, but as quickly as it was removed it reappeared, like the hypnotic, multiplying heads of Medusa. Everyone wanted to see Marimichael's naked swan dive and her new tattoo of the burning oil refinery.

As the video went viral, new protests erupted on the streets of the District of Columbia, New York City, Houston, Dallas, Atlanta, Chicago, Detroit, Los Angeles and other major cities.

Photoshopped pictures of Marimichael's dive appeared on social media. Instead of a high-dive platform, she now performed a swan dive from the torch of the Statue of

Liberty, the Seattle Space Needle, and church steeples. Some photos were strikingly composed and beautiful. One artist gave her wings. She suddenly appeared on T-shirts and posters.

Overnight she had become a cultural icon.

Sadly, she lay paralyzed in a military hospital, at an undisclosed location. The press had no access to her. She might as well have been held at Guantanamo Bay for all the rights she was given; in fact, some people thought she was being held there.

Liko was held at a federal detention center in West Virginia. The government had charged him with felony assault on a federal employee, the FBI man, and aiding and abetting the O'Briens. Consequently his bail was denied.

Climate crisis protestors now joined the violent protests. Large groups gathered in American cities, and sit-ins in local and state government offices occurred. In the nation's capital they shut down the city. In New York and other major cities, environmentalists protested outside the nation's major banks, shutting them down for funding fossil fuel companies.

Climate activists who tried to gather at the Lincoln Memorial for a march on the White House were immediately arrested and taken to Audi Field at Buzzard Point for processing. The new administration was determined to prevent a repeat of the Vietnam protest of 1967 that saw more than 100,000 protesters march from the Lincoln Memorial to the Pentagon, with the bloody carnage that then followed.

The *Post* headline read, "Military Rounds up Climate Protesters." The *New York Times* headline was also ominous: "Audi Field: New Federal Detention Center?"

Anarchists began to strike indiscriminately at the utility infrastructure across the country. Since they were seldom caught and arrested, their guerilla-style tactics began to disrupt the national power grid. They were joined in the streets by large numbers of disgruntled Americans who smashed and grabbed property from retail businesses. The poorest Americans were taking advantage of the national crisis to vent their anger at the wealthy "one percent;" unfortunately, their targets were usually small businesses in their home communities.

But worst of all, the terrorists, Marimichael's followers, continued their systematic attack against the fossil fuel industry with assassinations, bombings, and cyber-attacks. Their campaign against big oil escalated.

The COVID-19 pandemic had bankrupted some energy companies and weakened financial institutions that had backed big oil and the oil titans. During the pandemic, oil use had declined by more than thirty percent worldwide, and the associated sharp fall in price had damaged the US shale industry.

And now angry protestors attacked the fossil fuel industry. Their goal was to bankrupt not only the fossil fuel companies but also the institutions that funded them.

The stock market nosedived. The Federal Reserve used available tools to stop the economic panic and shore up the economy, but just like during the pandemic, their efforts had little or no immediate impact. After a week of volatile trading, the stock market crashed.

America's new president extended his declaration of national emergency relating to domestic terrorist acts. Congress shut down the US Capitol and all offices to the public. Non-essential federal employees were told to stay home.

During his presidency, Trump had deployed active duty soldiers to San Diego's San Ysidro border and El Paso's Paso del Norte Bridge during the coronavirus pandemic. Before that, he had ordered more than five thousand active duty troops to the United States–Mexico border to enforce his anti-immigration policies so his administration could surveil migrants.

The new president deployed America's military to all major transportation hubs: airports, seaports, train stations, bus stations, and even key intersections along highways. They joined National Guard forces previously ordered in place by the president.

Trump had declared a national emergency to give himself emergency powers during the coronavirus pandemic. President Clincher now took control of the nation's streets and neighborhoods, ordering a nationwide shut-down and a curfew between midnight and 6am.

Liko recalled being in Hawaii just after 9/11 when armed soldiers suddenly appeared at airports carrying automatic assault rifles. He remembered armed soldiers patrolling the small airport on the small island of Lāna'i. It had been spooky and disquieting. Now the presence of American military on the streets and at the nation's transportation hubs felt wrong, ominous.

According to *The Washington Post*, a memo from an undisclosed source showed that the president had also sought legal advice about suspending civil liberties—in particular, the right to *habeas corpus*. There was precedent for such a suspension: Abraham Lincoln and Congress had suspended the writ of *habeas corpus* during the Civil War.

President Clincher hinted that, if necessary, he would not only suspend *habeas corpus* but also other civil liberties, including "freedom of speech, or of the press; or the right of the people to peaceably assemble, and to petition the government for a redress of grievances." Newspapers reported that members of Clincher's administration were working with Congressional leaders on the necessary legislation. "Any authority or power that Congress grants the president will only be temporary, and it would have to be reviewed and renewed periodically," one Senator said. "Like the Patriot Act."

Across the country, people were scared and stayed in their homes. For days, no one went to work or school. No one went shopping, or to restaurants, or to gatherings such as sports events and concerts. The country became deathly quiet just like it had during the coronavirus pandemic. Liko thought, *It's like that scene from* The Day the Earth Stood Still, *but much more tense. Much more fear.*

Charla visited Liko at the detention center. She arrived dressed to kill, wearing sheer, puffy, elbow-length sleeves over a white camisole lining; tight black faux-leather pants; gray pointed boots; and a wide-brimmed black hat. The hat reminded Liko of Jack's red straw hat, except Charla's was felt and fashionable. He thought her outfit appeared feminine yet muscular.

"The guards didn't know what to make of my outfit," she laughed as she modeled for Liko, turning to her right and then to her left.

"Drama," he said. "You're all about drama."

She smiled. "Let me see your hand." He stretched out his hand wrapped in white gauze and bandages. She took it in both of her hands and held it gently.

During his short time in the detention center, Liko had had two surgeries, yet the hand remained mostly unusable.

Several fingers had one or more pins that realigned and reconnected metacarpal bones.

The force of his blow against the unyielding pool tiles had flattened his knuckles, literally. In the hospital, shortly after his first operation, he had foolishly removed the bandages from his hand and tried to make a fist. A serious mistake. Damaged bones shifted. A second operation had been required to reset the pins where bone had pulled away from bone.

And just last week, a pin in his thumb had bent. Liko found it surprising that such a small thing, a tiny pin in his thumb, could cause such a piercing, constant pain. He had trouble sleeping at night. Another surgery was now scheduled.

"I can't box anymore," he said. There was no complaint in his voice, even though he had enjoyed boxing. *I can't even grip a five-pound weight.*

"I'm sorry to hear that," Charla said.

"Don't be," Liko countered. "Maybe I'll take up dancing," he joked.

"I'd love to dance with you," she said with a smile.

"How have you been?"

"No complaints," she said. "Except the food in the hospital cafeteria could be better."

"How is he?"

"He came out of his coma yesterday." She smiled.

Liko had no doubt that she had been bedside, holding his hand, when he awoke. "He's going to be OK?"

"I think so. He's anxious to return to work and pick up his investigation of Dimer."

"He lives for his work, yeah?"

"He does," Charla said. "He thinks there is something off-center about Dimer's finances and he is determined to uncover it." Charla hesitated, seeming to decide whether to continue. "And something else, Liko."

"What?"

"He thinks it was Dimer who had him shot, not Marimichael."

"Really?"

"Yes. In most of the other assassinations, a drone was involved, and the rebels posted videos of the murders." She looked at Liko. "Was a drone present when you and Jack were attacked?"

"I never saw one." Liko now recalled how vehemently Marimichael had denied making the hit. And the rebels had no reason to shoot at Liko; nevertheless, he was certain that a shot had been fired at him, too. Liko now recalled how Gus had threatened him.

Danielle visited Liko in the detention facility weekly, bringing him paper copies of *The Washington Post* and, occasionally, *The New York Times*. She preferred her digital subscriptions, but she bought them especially for Liko because he was not allowed to have any electronic devices, including his iPhone or iPad. The papers became his lifeline to the events sweeping across the country.

Her first visit had been awkward and uneasy. After all, Liko had failed to get any pictures of the tattoo parlor other than a gloomy dark shot of a brick building. He also had failed to get any pictures of Marimichael, the rebels, or the FBI action. And he had failed to schedule an interview for her with Cody Long. Liko knew that he had failed Danielle miserably. Nevertheless, she forgave him. Yelled at him. Scolded him. And forgave him.

When Danielle visited Liko, she shared the progress she

was making on her series on the O'Briens. She pulled her reporter's notebook out of her back pocket and opened it.

"You going to interview me?"

She smiled and answered, "No. Why? Has something interesting happened since my last visit?"

"Nope," he said. "Nothing happens in here."

She pulled out a folded sheet of paper from the back of her notebook and handed it to him. "A copy of my last published piece on the O'Briens."

"Thank you," Liko said. He was genuinely happy to receive it. "I look forward to reading it."

Danielle smiled. "After Afghanistan, Marimichael took a job as an army recruiter, first in Maryland and then in Arlington, Virginia."

"She recruited soldiers?"

"Yes," Danielle nodded. "She also used her position to recruit like-minded environmental activists. She cherry-picked soldiers from the army's own recruitment database—both new recruits and retiring veterans."

"That's amazing," he said. "And unsettling."

"She did it for several years. The piece I'm working on now is eye-opening, too," Danielle added. "Once they released the identities of the kidnappers—the two rebels who died with Shane—I began tracking down their stories. And guess what?"

"What?" Liko said, "I'm on the edge of my seat."

"Their last service was in Syria. They were both members of a Special Forces group who worked with the SDF's elite counterterrorist unit. In Manbij, northeast of Aleppo, and in the Euphrates River Valley with the Kurds."

"The Kurds Trump betrayed?" Liko asked.

"Yes. Those Kurds. The rebels trained them in

surveillance and counterterrorism. And they trained together, like brothers."

"Let me guess," Liko said. "They also battled together, side by side."

"And trusted each other," Danielle said. "Would have died for each other."

"And then Trump betrayed them," Liko said.

"Yes," Danielle said. "And it set poorly with many of our men and women in the military. It went counter to what they believed in."

"Looks like our kidnappers were carrying some heavy baggage," Liko said.

"And Shane, too. He also battled with the Kurds."

"When? Where?"

"In 2014. The Battle of Kobani."

"Are there other rebels like them?" Liko said.

"Such as?"

"Other veterans who believe that the US betrayed them and their battle buddies."

Danielle and Liko looked at each other.

"Maybe the eco-terrorists should wear green berets," Liko joked.

"Maybe." However, neither Danielle nor Liko laughed. It wasn't funny.

After Jack was discharged from the hospital and then returned to work, he visited Liko in the federal detention center, too.

"Are you doing OK?" Jack asked. "How's your hand?"

"Improving," Liko said. "But it's a good thing I never wanted to play the piano." They both smiled, although Jack's smile was tinged with sadness.

Jack interlaced his fingers and set his hands on the top of the table. Liko thought he looked older, tired.

"How are you?" Liko asked.

"I lost a lot of weight. It's amazing how much muscle you lose just lying in a hospital bed. Charla has been great, though."

"She is special."

"I think so, too." Jack changed the subject. "I have some good news," he said. "Take a look at this headline." He pulled *The Washington Post* from its wrapping, opened it, and handed it to Liko.

The bold headline jumped off the page: "Delaware Papers Dumped!"

"Someone hacked into the database of a law firm in Delaware," Jack said. "They stole a treasure trove of financial and attorney-client information."

"Who has the data?" Liko asked.

"We do," Jack answered. "The hackers gave thousands of records to *The New York Times*, *The Washington Post*, and *The Guardian*. Fortunately, this time, they gave the FBI a copy, too."

"And?"

"It's evidence that some wealthy Americans are using shell companies for fraud and tax evasion. Foreigners, too. You'll soon see stories in the papers."

"Anything about Augustus Dimer or his Senator Nappe?"

"As a matter of fact, yes." Jack's voice rose with his excitement. "Look at the next page."

As Liko turned to the inside pages of the *Post*, he recalled that Danielle had jokingly asked him weeks earlier, as they drove from West Virginia to DC, if he had ever turned the page of a newspaper. Now he had.

Liko read the headline: "Senator Nappe's Hidden Wealth."

"And *The New York Times* is reporting payments between the fossil fuel industry and politicians." Jack slapped his thigh. "The payments were craftily hidden, but now they are coming to light."

"The data breach was huge," he said, continuing. "I hope to uncover something on Dimer, too."

"What will you do if you find something?" Liko asked.

"Investigate and indict."

The next day the charges against Liko were suddenly and mysteriously dropped. He found it disconcerting. He had assaulted an FBI agent and broken his nose.

However, Liko had pulled his punch. Surely the agent knew that; he must have seen it in Liko's eyes. Perhaps the FBI agent had influenced the decision to drop the charges? Maybe.

The charge that he had aided and abetted the O'Briens was bogus. At least, he thought it was. He must have led the FBI to the tattoo artist and inadvertently to Marimichael. *They followed me, somehow; how else could they have been at that swimming pool?*

Whatever the reason, the charges had been dropped and Liko was not going to argue or rock the boat. He was happy to be released. He walked out of the federal detention center with a weary smile on his face. He did not look back.

He did wonder, though, if Charla had confronted Jack about the microphone he had hidden in the abacus lamp stand. If she hadn't, he would.

29

Bike Ride

"Hi Charla!" Jack yelled.

Liko looked up and saw Jack waving at him and Charla as he walked quickly down the sidewalk to join them near the intersection of 21st St and M.

"Hi," Charla said, returning the greeting. It was amazing how much emotion Charla conveyed in just one syllable. Her greeting was bright, sharp, and pregnant with happiness.

Jack gave Liko a fist bump and then gave Charla an affectionate hug. She was dressed in tight-fitting blue jeans and a T-shirt that said "GOP—COWARDS, LIARS, IMMORAL." Her cap said "DUMP TRUMP!"

"That's my shirt," Liko told Jack, unable to resist commenting. "And my hat," he said, looking pointedly at Charla.

She adjusted her blonde ponytail so that it passed

through the back of the cap. "It's hot," she complained. "So much sun today!"

Liko shook his head. "I want it back. The shirt AND the hat. OK?"

"Sure," she said, pretending to pout.

The three of them turned their attention to the bicycles for rent. A dozen bikes stood upright in their kiosks. "How does this work?" Liko asked.

"Not sure," Jack answered.

Together they figured out how to download an app and unlock the bikes. Liko charged all three to his Mastercard.

"It's been years since I've ridden," Charla said.

"It's my first time," Jack said.

"Jack!" Charla exclaimed. "You're kidding, right?"

"No." His seat was too high and needed adjustment. He opened a black clasp and lowered the seat, wriggling it to the lowest setting. "How hard can it be?"

"Why didn't you say something?" She looked down the street for a taxi, but none was in sight. The National Guard and the military had regained control of the streets in DC. The rest of the country was still struggling, though.

"I'll be fine," Jack insisted.

"But you just got out of the hospital," Charla admonished him. "What would your doctor say?"

"Don't fall hard," he joked. When Charla didn't laugh, Jack quickly added, "I have great balance. And the first mile is downhill. And it's not far to the Tidal Basin."

"A couple of miles!" Charla said.

Nevertheless, Jack insisted, so Charla gave him a crash course in balancing, steering, and how to brake by squeezing the levers on both sides of the handlebars. "Remember," she said. "It's best not to go headfirst if you fall."

"Right," Jack agreed. "No flying over the handlebars."

Liko and Charla exchanged worried glances as Jack played with the gear shifter on the handlebar.

"Jack, that doesn't work unless you're moving," Charla said.

"Then let's go. Let's get this show on the road!"

Charla looked frustrated, worried, and mad. "Well then, Jack, why don't you just hop on and ride." A slight smile appeared on her face, "Like you did last night."

Jack blushed and Liko couldn't suppress a chuckle.

"Let's stay on the sidewalk," Liko suggested. It was a wide pedestrian walk, built to accommodate large families or groups of tourists.

Jack sat on the bike seat and pushed off hard with his right leg, starting at a fast pace. The bike glided down the wide sidewalk. He quickly placed both feet on the pedals. As the bike slowed it began to wobble side to side. He stretched out his leg to stop his fall, but the bike suddenly fell in the opposite direction. He lay on the concrete with the bike on top of him and his right leg high in the air.

"Are you OK?" Charla asked.

"Yep," Jack said. "I had no time to peddle."

Liko noted that Jack's "yep" conveyed as much apprehension as Charla's "hi" had boasted happiness.

"I thought it would be easier." Jack extricated himself and stood beside his bike.

"You're not a virgin anymore," Charla said, matter-of-factly.

Jack looked at her. Liko could see the surprise on his face.

"First bike ride," Charla said. "First fall."

The three of them laughed together.

At an intersection, waiting for a light to change, the

front wheel of Jack's bike suddenly turned inward and he fell onto the handlebars. He cussed but got up quickly.

"Are you OK?" This time there was fear in Charla's voice.

"Yes. I'm fine," Jack said. "The bike must not be balanced."

They made it to the docking station at 17th St and Independence Ave, near the Tidal Basin, without further falls and no injury, miraculously. As they docked their bikes, the DC Circulator Blossom Bus drove by. The exterior of the bus was all pink, having been decorated for the Cherry Blossom Festival. An advertisement on the side of the bus read "Blossom into Spring."

From there they walked to the first trees.

Marines stood guard beside Martin Luther King's granite relief among the flowering cherry trees. Assault rifles hung from their necks and rested against their chests.

"Do you think the rumors are true?" Liko asked Jack.

"What rumors?"

"That protestors disappeared? That some were last seen at Audi Field?"

When Jack didn't answer, Liko rephrased the question: "What have you heard, Jack?"

"Rumors ... mostly."

It was the "mostly" that caught Liko's attention. The word hung in the air. Jack had never hedged his answers before.

Charla interrupted: "The trees are magnificent! This one has double flowers?"

"No way," Jack said.

Liko bit his tongue, deciding, reluctantly, to let the subject be changed. "Let me see."

Charla had pulled gently down a branch so that they could inspect the blossoms. Liko and Jack peered around Charla at the deep pink flowers.

"It is a double blossom," Jack said.

Liko now smelled a faint almond fragrance that filled the air. As they strolled along the memorial walkway, they discovered that most of the cherry trees were single white-pink blossoms with five petals. Trunks and branches were shiny gray.

The half dozen paddleboats that floated on the Tidal Basin between the Martin Luther King Jr. Memorial and the Jefferson Memorial reminded Liko of Zahi. Liko recalled sitting on a wooden park bench watching children in paddleboats on Maritauqua Lake when Zahi had wandered upon him.

He remembered Zahi wearing his trademark tweed coat, even though it was too warm and he was sweating. Liko had thought him an unusual guy. *If only I'd known.*

"You OK?" Charla asked.

"I was thinking about Maritauqua."

"I plan on never going back there," she said.

"Me too," Liko said, without hesitating. "Ready for the Jefferson Monument?"

"Sure." She took Jack's hand and offered her other hand to Liko. He held it affectionately. They began the walk along the Tidal Basin, admiring the hundreds of cherry trees, all in peak bloom. Liko thought Charla enjoyed being in the middle of things.

"My god!" Charla said. "He's huge." They had walked up the steps and now confronted the statue of Thomas Jefferson.

"Definitely larger than life," Jack said.

"Definitely," Liko concurred.

"I like your shirt, sister," a man said, walking up to Charla. "Please, for you." He handed her a sheet of paper, bowed slightly, and walked away.

"Cheeky," Jack said. He watched the man carefully as he walked away. The man did not turn or look back; instead, he distributed his propaganda to other tourists entering the rotunda.

"What does it say?" Liko asked.

Charla handed the leaflet to Jack, who cleared his voice and read: "I hold it that a little rebellion now and then is a good thing, and as necessary in the political world as storms in the physical."

"Propaganda," Liko said.

Jack smiled. "It's a quote from Jefferson. He wrote it to James Madison."

"Get out!" Charla said.

"No," Jack said. "I've heard it before. I believe Jefferson meant just what he said, too."

"A little revolution?" Liko said. "What's that?"

"I'm sure he wasn't promoting assassinations and anarchy," Jack said.

"So he wasn't advocating violent revolution, like the O'Briens?"

"I seriously doubt it," Jack said. "But there's more. Should I read it?"

"Yes," Charla said. "Please do."

Jack looked at Liko.

Liko nodded.

Jack continued to read from the leaflet: "Unsuccessful rebellions indeed generally establish the encroachments on the rights of the people which have produced them."

Jack paused, which gave Liko the opportunity to

consider what Jefferson meant when he wrote "encroachments on the rights of the people." Liko wasn't sure.

Jack continued: "An observation of this truth should render honest republican governors so mild in their punishment of rebellions, as not to discourage them too much. It is a medicine necessary for the sound health of government." And it then says, "Join us!"

Jack looked up from the leaflet and surveyed the memorial. Liko guessed that he was trying to locate the man who gave Charla the leaflet. He was gone.

"What do you think, Jack?" Liko asked.

"I think governments must be held accountable to the people, and occasionally our leaders must be pushed to do the right thing. I think the climate crisis necessitates a 'little rebellion.' But assassinations? No. Bombings? No. I think 'a little rebellion' means disruption without violence."

"I agree," Charla said.

"So do I," Liko said.

Jack turned the leaflet over. "Oh my."

"What?"

Charla and Liko stepped over to Jack and looked at the picture on the leaflet. Marimichael had been Photoshopped into the iconic picture of Che Guevara, the picture taken in 1960 by the Cuban photographer Alberto Korda.

"She looks good in a beret," Charla said. "It will become the fashion."

"Charla!?" Jack exclaimed in an admonishing tone. Charla's name echoed inside the monument.

"Her swan dive already appears on T-shirts and posters," Charla said.

Jack shook his head. Dismay filled his face.

It was a melancholy walk to the next memorial. The phrase "we need a little rebellion" keep looping over and over in Liko's mind. And so did Jack's phrase "disruption without violence."

Liko decided that he liked disruption without violence: occupying offices, blockading roads, unveiling banners on bridges, gluing yourself to entrances, and mass arrests. But no network of violence like the O'Briens' rebellion. *Pouring buckets of fake blood?* he wondered. *Yeah. That would be OK.*

When they reached Franklin Delano Roosevelt seated in his wheelchair with his dog nearby, Liko was feeling better, less frustrated.

He was glad to be with Charla and she was happy to be in Jack's company. Liko was happy for them. Genuinely happy. He wondered if the three of them would become friends. *I never would have imagined an FBI man for a friend?* And then he remembered that Jack had bugged his apartment. Liko still needed to confront him about that. But not now.

"Fuck it!" he mumbled.

"What?" Jack said.

"I'm sorry," Liko said. "I'm still struggling. Given all this chaos and rebellion, what are we to do?"

Jack pursed his lips but said nothing.

Liko gazed at FDR's wheelchair. "I didn't know he was ... crippled?"

"Polio." Jack walked up to FDR and placed his hand on his shoulder. "My physical weakness is atrial defibrillation."

Liko wasn't sure exactly what that was, but it sounded bad. Really bad.

"I had an operation as a baby," Jack said. "And I've had several since then. I've been rushed to the hospital a time or two."

"And you dance flamenco?" Charla asked, startled.

"That's WHY I dance," Jack answered. "Dancing strengthens my heart."

"Oh, Jack!" She walked over to him and slid her arm through his and pulled him close to her side. "I love you!"

"I love you, too," Jack said.

"And you Liko," Jack asked. "What is your physical weakness?"

"Claustrophobia," Liko said. "It strikes at the worst possible times." *Like diving at the bottom of a pitch-black quarry. Or expecting an earthquake to bring the dome of a cathedral crashing down on me. Or like self-isolating during the coronavirus pandemic.* That had been the most nerve-wracking and claustrophobic event: self-isolating for weeks and weeks in a studio apartment.

Liko smiled and said, "I don't think Danielle would mind me telling you. She's dyslexic."

"But she writes a blog," Charla said.

"And she wrote a column for a major newspaper," Liko added. "Who would have guessed?"

"Not me," Jack said. "I've read her posts about Marimichael, and they are not only well written, but also great journalism."

"She told me the guy who started BuzzFeed was dyslexic, too," Liko added.

"How about you, Charla?" Jack asked.

"Me?" She thought for a moment and then said, "I'm a drama junky."

All three laughed.

As they worked their way through FDR's memorial,

Liko found himself surrounded by images of the Great Depression. He began to understand the brilliance of FDR's New Deal. Nevertheless, when they reached the end of the memorial, he was again thinking about the current state of affairs.

"Do you think we need a Green New Deal?" he asked.

"It's a no-brainer," Charla said, and Jack agreed.

Since they were all three tired, and because they didn't want Jack to ride a bike back to the West End neighborhood—especially since the last half mile had a steep hill—they caught the Circulator Blossom Bus. It was a comfortable and nice ride.

Liko enjoyed seeing Charla and Jack seated together, side by side. *Must be nice,* he thought.

He wondered if love was the antidote to chaos, the cure for the uncertainty each of us faces. Chaos was ever present. Even within himself.

Will I ever experience love again? he wondered.

He thought about Marimichael and their passionate sex on her squeaky bed, their bodies and minds intertwined for a night; they had had their chance, but they had not discovered love. For them it had not worked out.

30

Nightmare and Tampered Photo

Liko's nightmare had the feeling of a Dickens event, but it was set in the District of Columbia, not Dickens' London. The black hearse proceeded slowly down Pennsylvania Avenue with the victims of the assassins following directly behind, their feet and legs shackled, dragging ball and chains.

Though they bore the gunshot wounds that killed them—some with head shots, others with gaping holes through their chests—Liko identified several of them. There was Derichenko, dressed in the same suit he wore on the evening that Liko had knocked him out during the dinner at Davos. And there were politicians that he recognized by what was left of their faces, although he now had to admit, he had never learned their names. *Why would I*, he wondered, breathing heavily in his sleep.

The hearse stopped. The driver's door opened and Marimichael stepped out. Shane stepped out on the passenger side. Fintan mysteriously appeared at the back of the hearse and opened the back door. Together the brothers slid the coffin out of the hearse, allowing the legs on the underside to drop to the asphalt.

Marimichael stepped forward and opened the casket.

Liko saw himself lying inside. Pale face. Pale hands folded over his black tuxedo. A red bowtie.

Marimichael stroked his cheek. Then, one by one the victims shuffled forward to view his prostrate body. Some lingered. The faceless were the worst. They stared down at him.

Derichenko stepped forward and said, "You—"

As he awoke, Liko groped to silence his iPhone. He sat up in bed.

"Hello?"

"Mr. Koholua?"

"Yes?" Liko was sweating.

"My name is Sid. Danielle asked me to call you."

"Yes?" Liko looked at the time displayed in the top corner of his iPhone: not quite 7am. Sid, whoever he was, was calling at a ridiculously early hour.

"She hired me to authenticate several photographs?"

"Photographs?" He felt the bedsheet beneath him. It was damp with sweat.

"Pictures taken by Cody Long."

At the mention of Cody's name, Liko fully awoke. "What is it you do, exactly, Sid?"

"I am a digital forensic technologist. Danielle hired me to screen photos before she uses them in her blog or her newspaper stories. I have worked with her for many years."

"And?"

"And?" There was a pause. "I also screen pictures before they are submitted as evidence in criminal and civil cases."

"No, no, that's not what I meant. What did you find?"

"Mr. Long's video shoot of Marimichael—her swan dive—is an unedited original. His award-winning nature photographs, the ones that he took while working for National Geographic and that are now on sale at Getty Images, are also unedited originals. However, several of his recent photographs are fakes."

"Fakes? Which ones?"

"Photos on the recruitment thumb drive."

"Which photos?"

"There were several, but Danielle wanted me to tell you about one in particular. She called it the tree house picture."

Liko recalled the tree house picture, but not the details. He recalled opening the thumb drive and looking at the picture a long time ago.

"Are you still there?"

"Yes," Liko said. "The treehouse picture is a fake?"

"It was altered."

"How so?"

"It's technical."

"I'm listening."

"Someone was Photoshopped into the picture, and someone was Photoshopped out."

Liko tried to remember where he had left the thumb drive. "Can you send the picture to me?"

"It should already be in your email."

"Danielle gave you my email address?"

"Yes."

"Just a minute." Liko quickly clicked into his mail on his iPhone and opened the attached picture. It was as he remembered, except now he recognized Cody Long as the boy at the foot of the treehouse. The birthmark on his face was now evident.

"Let me guess," Liko said. "Was it the boy under the treehouse who is looking up at the O'Briens? Is he the person Photoshopped into the picture?"

"Yes."

"Why would someone do that?"

"I can't answer that," Sid said.

"Perhaps he wanted to be seen as an early friend of the O'Briens," Liko suggested, answering his own question. "Are you sure about all this?"

"Yes."

"How do you know?"

"It's technical."

"I'm listening."

"Every digital photograph has metadata associated with it. The treehouse picture was taken on a sunny day, so a small aperture was used. It's like squinting in the sunlight so less light hits your eyes. That's kind of the way the camera lens works."

"I see."

"As a consequence, the O'Briens are in focus, but the background is not." He paused for effect. "Except for the boy. He is standing near the back of the treehouse. He is in focus, whereas everything else at the same distance is out of focus. This is evidence that the boy with the birthmark wasn't in the original picture."

"I see."

"This is also supported by the ISO setting of 100, which, like the small aperture, is often used outdoors on a sunny day." Again Sid paused as he prepared to make his point.

"Yet the boy's right eye is noisier than the rest of the picture."

"What do you mean by 'noisier'?"

"Speckles," Sid said. "Lots of speckles. Which means the boy was shot at a higher ISO than the rest of the photograph."

"I see," Liko said. He had absolutely no understanding of what Sid had just said, but Sid sounded confident.

"There is other evidence," Sid said. "The pattern of shading on both the ground and the side of the treehouse is not consistent between the boy and the rest of the picture. The lens flare shows more than one light source. In fact two."

"OK," Liko said. "Anything else?"

"The image shows double compression, which means it was resaved, probably after being altered by photo editing software. In other words, the first compression quality was higher than the second. And to reiterate: the pattern of pixels does not correlate between the boy and the rest of the photo."

"Uh-huh," Liko said. He was now completely lost.

"Any idea who was Photoshopped out?"

"No," Sid said. "I can't answer that. But perhaps you can."

"I don't see how," Liko said.

"The metadata contains the GPS—the location—where the original photo was taken."

"So we know where the picture was taken?"

"Yes," Sid said. "According to Danielle, the GPS coordinates match the home of an ornithologist named Ace."

"Wow." Liko was surprised.

"The metadata also reveals the camera make and model.

Whoever took the picture used a Nikon model D200. Does Ace own this kind of camera?"

"I don't know." *But I will find out.*

"The owner of the camera is likely to have the original photo."

"I see," Liko said, and this time he did understand.

"If you find the original photograph, you'll discover who was Photoshopped out."

"I understand that too," Liko said.

"I listed all the metadata that I could validate and I attached it to the email. If you find the original digital photograph, the metadata will line up."

"Thank you," Liko said.

Why didn't Danielle call me and tell me all this? he wondered.

He opened the attached list: image size 4.2MB, image ID ClubHouse_5385.JPG, lens model AF Nikkor 18-70mm f/3.5 – 4.5G IF-ED, focal length 29mm, ISO 100, F number f/3.8, and much more.

Amazing. Taking a picture will never be the same.

31

Observations of Purple Martins

The surface of the pond exploded as the fish struck the bass popper. Liko instinctively jerked the pole, setting the hook deep. He gripped the pole in both hands, pulling it towards his head. The tip of the pole bent, forming a half arc, bending the pole back on itself. *It's too thin*, Liko thought. *It's going to break.* He released the line, allowing it to spin free. Tension immediately returned to the pole and the tip bent again. *Damn! It's huge! A huge bass!*

He played out the line. Reeled it in. Played out more line. The fish, struggling, felt powerful, strong, huge.

When the fish exhausted itself, Liko reeled it slowly, carefully to the surface. Its whiskered face broke the surface and its dark body shone smooth and slick. *A catfish? Dammit!*

Liko dragged the catfish onto the bank. He knew the

fins were sharp, so he stepped on it, carefully, to hold it in place. It was at least two feet long, and fat.

Liko reached down, carefully, to unhook the fish. The mouth opened and Liko saw the bass popper deep in its gullet. He had no pliers. "Dammit!"

He cut the transparent fishing line close to the mouth and then reached his fingers deep inside until he felt the popper. He wiggled it, jerked it side to side, and pulled on it. He had wanted to catch and release, but he now feared that the catfish would not live. Liko wiggled and jerked, but the popper didn't budge. Frustrated, he jerked harder and it suddenly came loose. He pulled the bass popper out of the gullet and mouth of the catfish and held it in the sunlight. It didn't appear damaged; the metal and plastic were stronger than flesh and blood. He gently kicked the catfish back into the pond, knowing that it would probably not survive.

His emotions had gone from ecstatic to disappointed in minutes.

Liko re-tied the bass popper to the line and cast out far into the pond. Almost immediately he had another strike. This one felt different—a powerful fish, but not a deep diver. The fish raced in one direction, then another, and then the line slacked as it raced towards Liko. With time, this fish exhausted itself, too.

Liko reeled it in. It was a good-sized bass. The popper was in its upper jaw and easily removed. Liko admired its size and weight, and its brassy-green coloring and the black stripe along each flank. He released the bass back into the pond, gently, with both hands. He watched its tail swish side to side as it swam away.

During the next hour, as the sun rose to noon, Liko caught several more northern largemouth bass, releasing each one. He grew tired. Thirsty, he walked up to the farmhouse. He wondered if he should have brought a lunch.

He also wondered how he was going to ask Ace about the treehouse photo. Did Ace take the picture? And who had been Photoshopped out?

Ace had picked fresh vegetables from his greenhouse, all of which he now stir-fried in a large cast iron skillet on his gas stove. Liko felt like he was receiving a special treat, especially the sweet snap peas.

Ace heaped the vegetables on two plates and set them on the table. He poured a glass of white wine for himself and one for Liko.

"Have any ice cubes?" Liko asked.

"Yes." Ace took a regular glass from the cupboard and filled it with ice. "Water?"

"No thanks. It's for the wine. I like it ice cold."

Ace smiled and passed the glass filled with ice to Liko. He closed his eyes with his nose in his wine glass.

Liko looked at the tanned face, the many wrinkles at the corner of his eyes, the thinning, short gray hair.

A moment later Ace opened his eyes with a full, cheek-to-cheek smile. He took a sip of the oaked chardonnay. "How is the fishing?"

"Excellent. I caught several large bass and one huge catfish."

"I've been catching and releasing that monster for years. Destroyed several of my best poles."

Liko dropped ice cubes into his wine glass and swirled them. He raised his glass and clinked it with Ace's. He took a swallow; the wine was ice-cold and perfect—a full-bodied white with a hint of baked apple.

"I enjoyed the fishing very much, but I have another reason for my visit."

"I suspected that you did. But you're always welcome here, to fish, to work in the garden, to stroll along the river path."

"Thank you. When I lived in Hawaii, I built raised growing beds." Liko recalled preparing the heavy soil; turning-under a growth of tall, scraggly weeds; constructing four-by-eight-foot planters. "We filled them with topsoil and planted seeds and seedlings. I enjoyed working with my hands. I miss it. I could build you a composter, if you like."

Ace smiled. "First, tell me why you came today."

Liko took out his iPhone and showed Ace the tree house picture.

"I remember that picture well." Ace handed the iPhone back to Liko. "I took it years ago. I gave a copy to Marimichael."

"Take a closer look. Someone edited it." He turned the phone is his hand so Ace could see the picture again.

"Oh yes!" Ace exclaimed. "The boy at the base of the tree was Photoshopped in."

Ace looked at Liko. "I think I know why you're here. I still have the original. It's somewhere in the cloud. Let me get my Mac Book Pro and we'll take a look. Shall we?"

"That would be great," Liko said.

Ace took out his Mac Book Pro, logged into his iCloud account, and began searching archived photographs.

"Is this the picture?"

Liko was sitting shoulder to shoulder with Ace, sharing his laptop's screen. "Yes," Liko said. "That's it."

Ace pointed to the boy standing below the tree house. "You'll never guess who that chubby kid is."

"Who?"

"The future President of the United States."

"No kidding!" Liko stared at the picture. *Clincher? President Clincher?* "He and Marimichael were childhood friends?"

"He spent summers with his grandparents. Not far from here."

The chubby kid was holding something in his hand. Liko leaned in to look more closely. A cat? He was holding it by the tail. It looked limp, dead. In fact, it had no head. "What is he holding?"

"A cat."

"It doesn't have a head."

"Correct. That's Topaz, the neighbor's cat. Clincher decapitated him."

Liko wanted to verify what he was seeing, but he had no desire to zoom in. It was grisly. The whole scene was disturbing, repellent, nauseating. *What a horrid gift.* Clincher was bringing Marimichael a dead cat—and decapitated, too. No wonder Cody had altered the photo.

"Can I have a copy?"

"Sure," Ace said. "Here, why don't you type in your email address for me."

Liko typed his email address into Ace's laptop and then Ace sent him the picture. Liko heard the whoosh as the email with the photo of the President and Marimichael and her two brothers—and the headless cat—went from Ace's mailbox to his iPhone.

Liko pulled up the picture on his iPhone. He studied the smile on Clincher's face, he looked at the headless offering, and then he forwarded it to Danielle.

"Maybe you should send her a note, too?" Ace suggested. "To prepare her for the shock?"

"Yes. I should do that." Liko typed a message, hurriedly, warning her: "I forwarded a disturbing picture. Open with care."

"Have another glass of wine?"

Liko nodded and Ace refilled his glass. Liko plunked in two more ice cubes.

"Why are you helping us?"

"The truth is the truth," Ace said. "I have no fear of the truth. Neither should anyone else."

As they ate, their discussion returned to the pond and fishing.

"I think I killed your prize catfish," Liko said.

"I doubt it," Ace said. "Catfish are tough—that one in particular. I've caught and released Old Charlie a dozen times."

"Well, I hope Old Charlie lives."

"He probably will."

"I forgot how much I enjoy being outdoors in nature," Liko said. "I thought about being a marine biologist, and now look at me. I live in a steel and glass building, surrounded by hotels."

Ace laughed. "The unexpected intrudes on our lives, doesn't it?

Liko nodded.

"Marimichael also wanted to be a biologist—a naturalist. Does that surprise you? She was never happier than when she was observing nature. In fact, I helped her publish her first scientific paper: 'Observations on Purple Martins.' We built a martin house together, and I mounted it on the side of my barn."

"I didn't see a barn," Liko said.

"I tore it down a few years ago and built the greenhouse

in its place. Keeps me in hydroponic tomatoes during the winter and early season seedlings in the spring."

"And the snap peas are delicious. Thank you."

"You're welcome. It's nice to have someone appreciate them. Now, where were we?"

"You mounted a martin house on the side of your barn."

"Oh, yes. Marimichael observed the martins through binoculars, or I'd hold the ladder for her when she climbed up to peek in the nests to count eggs, observe the fledglings, and band the martins. She was published in the *Bulletin of the West Virginia Ornithological Society*. I helped her with the edits. I think I still have her field notebooks."

"If it's not too much trouble," Liko said, "may I see them?"

"Sure." Ace got up from the table and disappeared. He was gone a long time, but when he returned, he had a medium-sized cardboard box. On the side of the box, in black marker, was written "Martins/Marimichael."

Ace set it next to Liko on the dining room table with a thump. "I'll make us some coffee," he said.

"Wait – when did she do this study?"

"She started her observations in the eighth grade, and she published her paper during her senior year of high school."

Liko removed the lid to the box and set it on the table. He gazed inside. He wished Danielle was present.

He removed two three-ring notebooks. The first was labeled "Observations of Starlings." "I thought you said she studied purple martins?"

"I forgot about the starlings. She studied them, too." Ace stretched out his hand. "May I see that?"

Liko passed the starlings field notebook to Ace.

Liko read the title of the second notebook aloud,

"Observations of Purple Martins, by Marimichael O'Brien." He set it aside, curious to see what else was in the box.

He discovered a purple martin nest resting in a jar with a tight-fitting lid. He rotated it slowly in his hands, viewing the disheveled nest from all directions.

"They tend to fall apart when you remove them from the martin house," Ace explained.

Liko opened a small cardboard box and found three, small white martin eggs. A shoebox contained the skins of adults and fledglings.

Wow, she was really serious about this.

Ace removed a loose sheet of paper from the front of the starlings field book. He sat quietly reading it. When he came to the end, he said, "Liko, I think you should hear this."

Liko paused his exploration of the contents of the box and looked at Ace. "What is it?"

"'Feeding Habits of the Great Horned Owls, *Bubo virginianus.*'" Ace smiled sadly. "It's short. One page. I'll read it to you."

"'On February 26, I found a Great Horned Owl brooding two young on the north wall of a barn in West Virginia. The adult female was the only bird seen to feed or brood the young. Most of the food items brought to the nest were decapitated, especially the larger forms.'"

Ace paused and looked up from the page to Liko. "And then there is a table listing food items brought to the nest by the Great Horned Owl, from February 26 to April 6th. The items are:

'Birds

14 Domestic Pigeon, *Columba livia*

6 Meadowlark, *Sturnella sp.*

3 Starling, *Sturnus vulgaris*

2 English Sparrow, *Passer domesticus*

1 Tree Sparrow, *Spizella arborea*

1 Harris's Sparrow, *Zonoirichia querula*

1 American Woodcock, *Philohela minor*

5 Bronzed Grackle, *Quiscalus versicolor*

2 Coot, *Fulica Americana*

1 Pied-billed Grebe, *Podilymbus podiceps*

1 Purple Martin, *Progne subis*

Mammals

6 Rabbit, *Sylvilagus floridanus*'"

Ace continued, "On the back of the page Marimichael sketched an outline of the Great Horned Owl. It was a large, marvelous bird. She sat in the barn and observed it on the ledge for hours."

"May I see it?" Liko asked.

Ace handed the page to Liko. When he saw the outline of the owl, he inhaled sharply and felt his heartbeat increase: it was the symbol used by the rebels. "Can I share this paper with Danielle?"

Ace hesitated.

"I'll just take a picture and send it to her."

Ace nodded his approval.

"Thank you," Liko said. He set the handwritten page on the table and focused on the title, "Feeding Habits of the Great Horned Owls, *Bubo virginianus*." It appeared sharp. He zoomed in and out again, just to be sure. He wanted the handwriting tack sharp. The camera focused on the authors' names: Marimichael and Charles.

Liko felt his heart race. "Charles is a co-author?"

"Yes," Ace said. "He and Marimichael spent hours – days – together observing the owls."

Liko's hands shook, slightly, as he took pictures of the

front and back. He then handed the page back to Ace. Liko hoped that the picture was in focus because his fingers were trembling.

He wondered if the handwriting was Marimichael's or the president's. He turned to look at Ace, held his gaze. "It's *his* handwriting?"

"Yes." Ace carefully folded the document and then returned it to the front of the notebook.

Liko opened the "Observations of Purple Martins, by Marimichael O'Brien" field notebook.

"OK, now I'll get the coffee going," Ace said. He disappeared into the kitchen. Liko took a deep breath and started to read:

FIRST YEAR

April 6

The following incident occurred during a period of inattentiveness with adult Martins absent 5:14pm to 5:20pm. Two pairs of House Sparrows alighted on the box roof. They appeared to be highly excited and/or heavily parasitized. At 5:17pm, a male House Sparrow, closely followed by another, carried a Martin egg to the ground. The first returned to the box and carried a second egg to the garden path 25 feet from the box. The egg was punctured but the liquid contents were not eaten.

5:54pm: Saw two adults fighting (?) in the air. They came together twice and rose high in the air and then separated.

6:45pm: Two adult males fell from box to shed roof 14 feet below while fighting. Both flew up and returned to opposite sides of the box. Two adult males fell to roof, rolled and tumbled 15 feet down the roof, fell off into a small bush, still held each other for a half minute. One adult broke loose and flew away. The other remained motionless in branches of bush with wings spread for 20 seconds. Flew back to box in a slightly erratic flight to place where fight started. Fight probably lasted two minutes. Several less intense fights are occurring.

7:06pm: As far as I can determine there are 7 males and 2 females in and around the box. Two pairs seem to be mated. Males are continually fighting and chattering a great deal.

7:16pm: Getting dark. Fights increasing in frequency but not in intensity. An adult male keeps three holes on west side of box all to himself driving out all intruders.

Marimichael had drawn diagrams that labeled the holes in the martin house—east, west, north and south—so she could count and record the number of birds. Her observations continued:

July

Martins drinking pond water. Adults swooping over water and drinking on the wing. Adult female dipped into water twice, circled, dipped into water a third time. Another adult female dipped into water eleven times. Racing dive – skim across water – almost stops flight. Adult female landed in top-most branches of elm tree. Ruffling feathers. Flew away in wide circle.

August 27

Nest 1. Bits of elm leaves, pieces of stem about 3 inches long, partially burnt piece of paper. Approximately ½ silt or fine dust in bottom of compartment. Pieces of elytra or wing coverings of beetles. Bark 1 inch long, ½ inch wide. Intestinal fluke (?) an inch long with red head, body color yellow. Wiggles but doesn't move about very fast. Numerous rapid running lice and small red mites in liter. Pieces of elm twigs 1 ½ inches long. Willow leaf. Manure in corners of compartment.

During the year over 50 percent of the nestlings perished in or near the box.

SECOND YEAR

July 14

Nestling found dead in possession of neighbor's cat next farm north.

THIRD YEAR

The cat caught a nestling. I saved this bird which was badly frightened and placed it in a cage where it lay on one side. An hour or so later this nestling was still on its side. I raised the bird by one wing and released it. The bird seems OK but was badly frightened by the cat. I banded the bird and released it. It flew about 200 feet and landed on a window awning. An adult martin followed the nestling but didn't alight on the awning.

June 22

7:00pm: Temp. 85 degrees F, south breeze, sky mostly clear with a few very high, scattered clouds. It rained last night and this morning. I have set up the ladders and readied my banding equipment for tonite's banding operation. I don't expect to find as many martins to band this year as last years record total of 50 birds in the 3-floor box.

8:15pm: A neighbor's cat climbed up the ladder leading to the barn roof. The martins were alarmed but I quickly chased the cat.

8:17pm: The birds have returned to the box.

8:20pm: It is getting dark. Bats are flying.

8:45pm: Boarded up box. Banded the birds.

10:20pm: Secured for the night.

June 23

The cat Topaz came over back fence into Ace's yard. Martins immediately flew from box giving alarm calls. The cat is watching. Six martins quickly congregated on wire. Cat in garden, then walked off.

June 28

5:00pm: Nest no. 18 has young birds! They are seen through binoculars and are in the SW corner of the compartment. The lining of the nestlings throat is a bright yellow with margins paler.

July 15

8:53pm: I started to board up the box at 8:30pm and just finished. It was dusk and birds were moving in the compartments. Not a single bird has escaped. From touching the box, I now have a lot of bird mites running over my ears, arms.

Marimichael then described the contents of each nest and her observations as she banded each fledgling on the left leg:

Struggled a little. Squawked! All birds examined have mites.

Topaz killed a nestling. I made it into a study skin.

FOURTH YEAR

February

Now that the martin box is up we can look forward to a new season and reminiscence of the past. This is a fine box and one to be proud of – 24 compartments on three levels.

June 23

4:40pm: I found a nestling just west of the barn below the martin box. It is dead but not stiff. The feathers have not pushed through the skin. The eyes have not opened but appear as slits. This nestling appears to have been pecked on the head and lower back since the skin is broken. The mouth and tongue are an orange-yellow color. The edge of the mandibles are white, tips of bill blue-violet. Some of the visceral organs can be seen through the skin of the belly region. Alula clearly distinguished. Color of skin on underside, legs, feet, side of neck a light ivory. Neck and head region almost as long as body. Belly distended and round. Made into a study skin.

FIFTH YEAR

August 22

The box was dismantled this afternoon.

Nest no. 2: Martin nest. Depression in center at rear of compartment but filled with scales of feather shafts, broken pieces of green elms leaves and a bean pod. Nest largely of walnut twigs, a few dried grasses broken bit of glass, small (1/2 inch long) shell which has a bright red back with ivory undersides, few remains of butterfly, hard parts of beetles.

Unusual nest items include a blue razor blade, a twisted empty pack of cigarettes. The latter is wrapped in cellophane, white paper and tinfoil. A scrap of burned paper that is scorched around the edges, a burnt wooden matchstick, a rusty nail.

After he finished reading the field notes, Liko read the published scientific paper, "Observations on Purple Martins," by Marimichael O'Brien. The opening paragraph was:

For five seasons, purple martins, *Progne subis subis*, were studied at a colony in West Virginia. This paper presents the results of observations on the nesting cycle, populations, and interspecific relationships with house sparrows, *Passer domesticus domesticus*. Individual birds were identified by bands and differences in plumage.

After finishing the paper, Liko quickly looked again at

the paper's sub-headings, re-reading the first sentence of each. One caught his attention: "Bandings. Seventy-six adults were banded." Liko set Marimichael's published paper on top of her field notebook. *Impressive.*

When Liko got back to his condo, he received a call from Danielle. "I received the picture!"

"That's great," Liko said.

"I've forwarded it to Sid for authentication."

"I'm sure it's the original," Liko said. "Did you get the owl study?"

"Yes."

"Ace said it's in Clincher's handwriting. He spent several summers with his grandparents. Looks like he was friends with the O'Briens."

"Great work, Liko!" And then Danielle volunteered: "I may have an interview with Marimichael. I'm working with her lawyers. Trying to schedule it now."

"She's still in a military hospital?"

"Yes. And she's paralyzed from the neck down. A respirator breathes for her and she is fed through a stomach tube."

"How can she talk if she is on a respirator?"

"That's a good question Liko. I asked her lawyer the same thing. The respirator is connected to her trachea, but some air still passes through her vocal cords. She is also strong enough to breath for short periods of time without the respirator. So we'll see."

Liko recalled watching Christopher Reeves, who had broken his neck in a horse accident, speaking at the Academy Awards, his voice soft, each word and phrase carefully measured.

"How is she mentally?"

"Her lawyer says that she is still fighting," Danielle answered. "The government will not honor her end-of-life health care directive."

"Why?"

"The Justice Department insists that her crimes preempt her health care directive."

"So they are refusing to honor it."

"That is correct," Danielle said.

"They should let her die," Liko said. "That's what I think."

"That may be why her lawyer is granting my request for an interview. I think Marimichael wants to shine a spotlight on the issue."

"You mean the right to die?"

"Yes."

"So she has her agenda..."

"And we have ours," Danielle said.

32

Seizing Power

Investigations into the Delaware Papers exposed Senator Nappe. The leaked financial records provided ample evidence to indict him on criminal tax violations, similar to Al Capone and Paul Manafort.

People were angry and frustrated, and the public shaming came swiftly on social media. The assault was rude and crass and lacked civility. It was, though, effective: the day after his indictment, disgraced, Senator Nappe lost his coveted position in the Senate.

Environmentalists cheered on the steps of the United States Capitol.

Senator Nappe's replacement allowed the climate crisis legislation—the Carbon Pricing Bill—to move to the Senate floor for a vote. The bill Senator Nappe had throttled for years easily passed.

The bill moved to President Clincher's desk, where he signed it into law, with great ceremony and fanfare in the

Oval Office. He signed his full name, Charles Chadworth Clincher, one letter at a time, gifting the 24 prized pens to his supporters. Liko recalled how Nancy Pelosi, the former Speaker of the House, had similarly signed impeachment articles against President Trump.

What Jack had hoped for occurred: democracy worked, and meaningful climate crisis legislation became law. Liko was happy, too. He imagined that Shane would also have been pleased: passage of climate crisis legislation had been his main goal.

During the following month, facing a lengthy jail sentence, Senator Nappe turned witness for the prosecution. As part of a plea bargain for a reduced sentence, he gave damaging evidence against the billionaire Augustus Dimer.

Dimer's mansion was once again raided, but this time by the FBI, not his private mercenaries. Jack carried the search warrant.

Dimer's son Gus was shot and killed while trying to evade the police. Jack found Augustus after he put a gun in his mouth and pulled the trigger. Their deaths were petty and selfish, just like their lives.

The administration rebranded the Climate Crisis Act as the Clincher Act. The president claimed the legislation as a personal victory for himself and his party, and he lauded it, every chance he got, both domestically and internationally.

During his first State of the Union address, he highlighted the Clincher Act, referring to it over and over again. He explained the scientific evidence for the climate crisis to the joint session of the United States Congress and to the American public in a no-nonsense, matter-of-fact message. By the end of his address, Americans finally understood they faced a life-threatening crisis. A crisis much worse than the COVID-19 virus. Whereas the coronavirus had threatened mankind, the climate crisis threatened all life, all species on earth.

President Clincher's message was broadcast around the world. "Once again we are at war with an invisible enemy. We fought and won the war against COVID-19. We must now fight and win the war against greenhouse gases. We must not fail. We will not fail."

The president assured Americans he understood the mandate they had given him. He assured his followers he would end the climate crisis. He promised to act boldly and decisively on their behalf. He called on all Americans to help each other—their neighbors, their children, and their children's children.

Americans' response was phenomenal. Overnight, people began attending town hall meetings. They demanded that their congressional members take immediate action on climate issues. Congressional members who failed to listen faced large numbers of angry constituents.

Newspapers printed front-page stories about the fossil fuel industry's influence in government, and their lies. Teams of journalists investigated the congressional super PACs and the Senate Leadership Fund.

Newspaper stories educated the public about the nexus between the fossil fuel industry's influence in politics and the gridlock in government. Americans learned that "freedom of speech" had become synonymous with "corporate money." Americans learned that fossil fuel companies, in particular, used corporate money to make political donations and to fund politicians. Americans learned that corporate money enabled oil barons like Augustus Dimer to buy politicians, to manipulate the marketplace, and to control Congress.

A bold headline appeared on the front page of *The Washington Post*: "AMERICANS DEMAND CAMPAIGN FINANCE REFORM!" In response, the president gave a speech from the Oval Office asking Congress to kill the filibuster. His words reverberated: "Down with the filibuster. Up with democracy! Down with dark money. Up with transparency!"

The Clincher Act had been the tipping point.

Meanwhile, Danielle continued to work on her article that connected President Clincher and the O'Briens. Her interview with Marimichael had been postponed and then canceled and then rescheduled. Danielle was anxious to publish, but she thought the interview was critical to her article's success.

Over the last month or so, the article had evolved, and so had Danielle's sources. Danielle said that she now had enough material for several articles.

"I'm in discussion with *The New Yorker* magazine for a possible series," Danielle told Liko. He could hear the pride in her voice. "Marimichael's relationship with President Clincher would be the lead story."

"Please be careful," he told her. "Let me know if you need anything."

33

Meteor

Polls showed that Americans' confidence in their government was rising steadily, and since democracy appeared to be healing itself, Liko imagined that Jack would be happy. But when Liko had last seen him, Jack seemed preoccupied, pensive, worried.

"He hasn't been himself lately," Charla said.

"Maybe he's just depressed," Liko said. "I mean, he spent years investigating Augustus Dimer, and then to have the case end so suddenly, with a suicide. And Gus dead, too. How do you deal with that?"

"You know that Jack witnessed the suicide? He arrived just before Dimer killed himself."

"I didn't know that."

Liko recalled the scene in *The Shawshank Redemption* when the warden was cornered in his office by the state police, and he did the same thing, blew out his brains. Movie suicides were one thing, but real-life suicides were something else. How do survivors live with the memories? And even more, how does a witness live with the image? Was that Jack's problem?

"Jack asked me not to tell anyone, but I'm worried about him and I trust you."

"It's OK, Charla, I understand. I won't bring it up."

In fact, Liko had his own secrets. He had told no one, not even Charla, about the clubhouse photo and Danielle's investigation. The article would be a major scoop for Danielle, and he didn't want to muddle things like he had at the tattoo parlor.

He was still angry at himself about that. *I didn't get a single picture of Marimichael, the rebels, or the FBI raid.* That failure, like so many others, looped in his mind all the time.

"I'll see if Jack wants to go on a hike," Liko suggested. "I heard about a trail not far from here that follows a creek. Maybe he'd be interested?"

"That would be great. But nothing strenuous. Remember, he has a heart condition."

"I'll be careful."

The next day, when Jack stopped by the condo after work to visit Charla, she told him that she was under the weather. "I need to lie down for a while. Maybe you could visit with Liko while I take a short nap?"

"Sure," Jack said. He picked up *The Wall Street Journal* from the coffee table. Zahi's subscription had not run out and the paper still arrived each morning. Liko had already decided not to renew the paper; he had no interest in financial or business news, and the editorials, in his opinion, sucked.

"Want to explore the trail along Rock Creek?" Liko asked.

"Oh, I dunno. I can read the paper while she naps."

"It's a nice day."

Jack shrugged.

"I could use the company."

Jack looked over the top of the *WSJ* at Liko. "OK, but just a short walk."

"Sure," Liko agreed. "We can do that."

Jack folded the paper and slapped it against his knee. "I'm ready."

They walked down 23rd Street to P Street and began the descent down forty-plus steps to a grassy field next to a creek. The steps were a steep combination of earth and railroad ties. Jack gripped the iron pipe handrail and took each step carefully. At the beginning of their walk, they managed a bit of small talk but soon slipped into a companionable silence, each pausing to quietly point out something to share with the other.

At the bottom of the steps they followed a short, well-worn dirt trail through the grassy field to the bank overlooking Rock Creek. They paused at the top of the bank. To their right, the creek flowed under the Lauzun's Legion Bridge and then meandered around a bend, disappearing behind the massive stone bridge.

Beneath the bridge, a lone blue heron stood on a gravel bar in the middle of the creek hunting fish near a riffle. Liko had seen blue herons before, but their large size always surprised him. After a lingering pause to watch the elegant bird, they decided to walk in the opposite direction so they wouldn't disturb it.

They followed the dirt footpath along the top of the creek bank until they entered a riparian woods. After that, the footpath climbed uphill, gradually, while staying parallel to the creek. The ascent was steep. Liko stopped so Jack could catch his breath.

To their right, the steep bank now fell fifty feet to the creek below. To their left, the bank climbed more than a hundred feet to a plateau. Although out of view, Liko knew that the plateau held a playground, softball field, pubic swimming pool, and a dog park, because he had studied their route on Google Maps in preparation for the hike. *Should I ask Jack about bugging the abacus?* Liko decided to save that question for another day, and to simply enjoy his walk with his troubled friend.

The dirt trail now slanted like the rest of the steep bank, making it difficult to walk—not because it was wet and slippery, but because it was dry and powdery, a fine clay that rose up in a puff with each step. Liko's white Adidas were now covered in a fine brown powder.

Virginia bluebells grew sparsely on the steep slopes, shaded by the trunks and leafless branches of elms and oaks and the tangled undergrowth. Their trumpet-shaped blossoms were a medium blue; Liko thought their delicate color and drooping blossoms were beautiful. He thought about pointing them out to Jack, but he seemed lost in thought so Liko admired the flowers privately. A dark-green English Ivy, an invasive species, appeared to be thriving, growing in large patches on the steep banks. A scrawny redbud tree was in full bloom.

Together they gazed down the steep bank. The creek was braided and shallow, but Liko guessed that its calm appearance was deceiving. There could be areas of deeper, faster water. And the depth of water was unknown.

Liko's eyes journeyed across the creek and up the opposite bank. Cars motored down Rock Creek and

Potomac Parkway, and bicyclists passed joggers on a wide bike path. *Next time I'll bike*, he thought. *Charla would like that.* He recalled Jack's poor biking skills and laughed, silently.

They continued their hike, passing large trees that had succumbed to the steep banks and gravity. Oaks bowed towards the creek, their roots under tremendous stress. Some trunks had ripped open; others had fallen into the creek, their crowns partly submerged. Fallen trees that crossed the dirt path had been chain sawed.

They came upon a middle-aged woman pulling up English Ivy. She was on her hands and knees on the steep hillside below them. Her task was impossible; nevertheless, she worked with focused determination, her gloved hands attacking the invasive ivy. Patches of the ivy were tinted blue and had withered. Liko guessed that the ivy had been sprayed with a pesticide that contained a blue dye.

"I never would have imagined a wild, riparian area like this in the heart of DC," Liko ventured. "Did you know this was here?"

"This footpath? No. But I have visited Rock Creek Park. It's farther north. It's large, almost two thousand acres."

"That would make it twice the size of Central Park," Liko said.

"Would it?" Jack said. "I've never been to Central Park."

"Have you started a new case?" Liko asked. He tried to sound positive and supportive.

"No, we're still wrapping up Augustus Dimer." Jack's fragile good mood quietened.

Again Liko tried to think of something positive. "I imagine it's better working for Clincher than it was for

Danson – and especially for the Trump administration, yeah?"

Jack stopped and stared at Liko. The expression on his face was one of surprise mixed with sadness and worry and fear.

Liko had thought the question would be an easy one—what Danielle called a softball question. Jack loved his work and Liko thought he would be happy to talk about it. Instead, he had accidently hit a nerve.

"I thought things were going well," Liko said. "At least with climate mitigation."

"To be honest," Jack said, "it sucks. It's worse than Trump."

"Really? That surprises me."

"Trump was incompetent and knew nothing about running a bureaucracy. Neither did the deplorables he brought into his administration. Clincher is different."

"How so?"

"What I'm about to tell you is my personal opinion. You understand, Liko? I'm not speaking on behalf of the FBI or as an employee of the FBI. OK?"

"I understand." He tried to make a joke: "What's said on the Rock Creek Trail stays on the Rock Creek Trail." Jack, though, frowned.

"I'm afraid that Clincher is a demagogue, a manipulator, a despot. The English Civil War produced Cromwell. The French produced Robespierre. I'm afraid the O'Briens have produced Clincher."

"But I thought things were going well! The Clincher Act. The end of the filibuster, and corporate and dark money influence in Congress. The weekly presidential briefings."

"I wish I shared your optimism."

"You don't?"

"No. Every day I see Clincher and his extremists—yes, extremists—rising."

"Rising?"

"Consolidating his power. He is promoting his climate agenda at the expense of our rights, our freedoms, our American values—and our democracy."

"But he's making progress. Things are getting better."

"Yes. That is true. Things are getting better. The economy too. But only in the short term."

"So what are you worried about?"

"When Mussolini took power in Italy, he succeeded at first, too. The Italian government functioned better, the economy improved, and he made good decisions. The Italians were happy. But it wasn't long before it all turned bad."

Jack stopped walking and turned to face Liko. "We are still under a national emergency, Liko. We still have a nightly curfew. The National Guard still controls the streets and the military controls all transportation. In the meantime, I am afraid that Clincher is consolidating, concentrating his power."

"But all presidents do that during a time of emergency," Liko said. "Don't they?" He lost his footing and slid a few feet down the bank towards the river. They were on a steep embankment. He climbed back to the footpath.

"But how long will this emergency last?" Jack asked. "He has extended it once."

As they continued along the trail, Liko noticed that they were surrounded by what first appeared to be squirrels' nests. Brown leaves appeared nested in the crooks of trees and underbrush. It was strange, though, that the nests were all so small. And then he understood.

The climbing became difficult and they stopped talking. Liko started to worry about Jack, and whether the trail was now too difficult. He let Jack take the lead, so Jack could set a comfortable pace for himself. He immediately realized that he had made a mistake; Jack liked a challenge and he had increased, not slowed, their pace.

"Wait a minute, Jack," Liko said. "Check out these clumps of leaves and debris."

Jack stopped and looked at the clumps.

"At first I thought they were squirrel nests," Liko said. "They're not."

"Then what are they?"

"This is how high the river has flooded. And from the heights of the debris, the floodwater was way above our heads."

"We would be under water?" Jack asked.

"Yes." *Not only that,* Liko thought. *When the river jumps its banks, it is impossible to swim against the current.* Liko looked at the sky. There were some clouds in the sky, yet he saw nothing ominous; nevertheless, the path no longer felt as safe.

Liko wanted to talk to Jack about Dimer's suicide, but he didn't. He wanted to tell him about Danielle's investigation, and her pending series about Marimichael, the O'Briens and President Clincher, but he didn't. And now he wanted to talk to Jack about the Clincher administration.

"What's the matter?" Jack asked Liko. "Something on your mind?"

34

Freedom of the Press

A whistleblower in the White House came forward claiming that President Clincher had abused his office by ordering military cyber-security experts who had been assigned to the White House to hack into the law firm of Wells, Nickel, and Phineas to steal the Delaware Papers. The president's men vociferously denied the accusation.

The White House press secretary infamously said: "There is no whistleblower. Just someone with an agenda against Charles Clincher. Get over it." The press secretary labeled the whistleblower a spy.

The whistleblower's name was accidentally released and she was identified as a low-level CIA operative assigned to the White House. That evening, on her way home from work, she drove off the road into the Potomac River, an apparent suicide. After her death, no one in the White

House would comment further about the hacking. All they would say was, "Our heart goes out to the family of the whistleblower. There is an ongoing investigation."

Danielle was not intimidated. Her most recent blog post had shed light on the rebels: who they were, where they came from, what they wanted, how they were organized, why they were so violent, and speculating on when the violence would stop.

Her blog was well received by her peers, and provided cannon fodder for other investigative reporters. As rebels died, or were captured, or their identities became known, journalists began exploring their individual stories, interviewing family members, neighbors, and in some cases former battle buddies. Why had these men and women betrayed their country?

The day after the death of the whistleblower, Danielle announced her forthcoming feature story in *The New Yorker*: "The O'Briens and President Clincher."

"You're poking a hornet's nest with a very short stick!" Liko said, talking to her on the phone. "Why?"

"Liko, they already know what I'm doing. What I'm working on." She paused to catch her breath. She sounded angry. "Wake up! Do you honestly think we have any privacy anymore? Any of us?"

Liko thought about the FBI bugging his home and tracking his whereabouts. Maybe she was right.

He had read in the newspapers about the federal government tracking immigrants during the coronavirus pandemic. Were all Americans now under surveillance, including Danielle and himself? Had Edward Snowden's nightmare finally come true?

"The FBI raided Sid's home," Danielle added. "They took his computers and all of his photo files. He's traumatized. Fortunately, he already authenticated the photograph and hid copies for me.

"When is your interview with Marimichael?"

"Four days from now, next Tuesday. Right after lunch."

"Need a driver?"

"No, Liko. The conditions for the interview are strict. No one is allowed but me. No driver. No photographer. Just me, my notebook and pencil, and my tape recorder."

"Good luck."

On Monday, the day before Danielle's interview with Marimichael, President Charles Clincher held a press conference in the Rose Garden. The weather was overcast and windy. The roses were not blooming.

The president's standing in the public polls was exceptionally high for someone who had entered the public limelight because of an assassination. Perhaps his high standing was because he had behaved with great discipline, like a general, not like a politician.

Americans admired Charles Clincher. Trump had filled his administration with sycophants, novices, yes-men, and businessmen prone to corruption and malfeasance. Clincher recruited men and women committed to saving the earth from the climate crisis. Whereas Trump loyalists showed blind fealty to Trump, Clincher's supporters were devoted to preventing the sixth extinction.

Standing next to Clincher in the Rose Garden were the Speaker of the House of Representatives to his right and the Chief Justice of the Supreme Court to his left. The Secretary of the Treasury was also stood with the president.

A few minutes into his prepared notes, President Clincher said, "I would like to share a quote from former

Vice President Joe Biden. During a Democratic debate in 2019 he said, "The first thing that happened when President Obama and I were elected? We went over to what they call—and some of you are military women and men—over to the tank in the Pentagon, sat down and got the briefing on the greatest danger facing our security. You know what they told us it was? The military? Climate change. Climate change. The single greatest concern for war and disruption in the world. Short of a nuclear exchange, immediately. So where are we?"

President Clincher looked up and addressed the camera. "Today I will tell you where we are. And let me be perfectly clear. Perfectly clear. Today, I declared a national emergency to combat the climate crisis. Make no mistake: the climate crisis is an existential threat to humanity and all other life on our planet.

"That is why, today, I have signed a new proclamation: The Declaration of a National Emergency and Invocation of Emergency Authority Relating to the Climate Crisis. You will be hearing more about this in the days and weeks and months ahead.

"In conjunction with this proclamation, today I am sending Congress a tax proposal—the Paris Hilton Climate Tax—that will fund a national Climate Crisis Campaign. To the wealthy one percent of Americans, I say, enough is enough! To the rest of Americans I say, plan for victory!

"The single strategic goal of the Climate Crisis Campaign is to ensure that our industries, our universities, our laboratories, and our businesses shall all be at a competitive advantage with the rest of the world as we tackle the greatest existential threat of our lives, the climate crisis. US industries must be competitive worldwide. US universities and laboratories must be energy-innovative. And US businesses must develop and produce the next generation of energy renewables.

"Make no mistake: We are in a clean energy race. We are competing, side by side, with all the nations of the world. And I say to all Americans, the Climate Crisis Campaign will ensure that the United States of America wins the clean energy race."

President Clincher turned to the next page of his speech. A camera zoomed onto the page and revealed that the notes were handwritten, ostensibly in the president's own hand. The handwriting was neat and clean and confident.

"One half of global emissions can be traced to just 25 corporations and state-owned entities. To the handful of petro-states, and to the oil companies that continue to produce fossil fuel, I say: Enough is enough. Today Americans are putting you on notice. The fossil fuels that are in the ground shall remain in the ground.

"Together we will stop the sixth extinction of humans and other animals. Together we will save the earth for our children and our children's children. Together we will preserve the earth for future generations.

"The Speaker of the House of Representatives and the Chief Justice of the United States are here this morning standing in solidarity with us. I have asked them to say a few words. Afterwards, the Secretary of the Treasury and I will answer your questions.

"Speaker Stewart." The president warmly shook the hand of the Speaker of the House as he stepped to the podium.

Charla turned to Liko. "It's not politics as usual."

35

An Interview Gone Bad

"She's dead, Liko," Danielle said. "Marimichael is dead. Her wish to die was granted earlier this morning."

"What?" Liko said. "What about your interview?"

"I talked to her attorney last night and he gave no indication that she would be taken off life support."

"Are you OK?"

"No, Liko. I'm furious." Her voice had risen an octave. "They took my iWatch and tape recorder and all my jewelry. And they strip searched me!"

"Where are you now?"

"In the jail lobby. As soon as they return my stuff, I'm out of here."

"Where can we meet?" Liko asked. "What's the best place for you?"

"The Cosmos Club. In forty minutes."

"I'll see you there," Liko said. "Watch your back."

As Liko walked to the Cosmos Club, his mind raced. He was worried for Danielle's safety. He also felt sad for Marimichael. His mind was racing because he believed that she had been silenced, murdered. Did someone take her off life support not to honor her wishes, but to hide the truth? If he was correct, then Danielle was in danger.

It took Liko less than fifteen minutes to walk to the Club. The doorman was not present, so Liko pushed the round intercom button.

"May I help you?"

"I am a guest of Danielle Queen. She will be meeting me here."

Hearing an electronic buzz, Liko pulled open the front door and let himself in. He had spotted the camera, so he knew that he was being watched. He proceeded directly to the couch in front of the large, decorative fireplace in the lobby and sat down.

When the doorman arrived, they exchanged nods. Liko sat and waited, gazing into the flames. His mind drifted. When he checked the time, more than forty minutes had passed.

Where was she? He tried to call but she did not answer her iPhone or iWatch.

The doorman gave him an admonishing look. "Phones may not be used within the club, except in designated areas. Or you can step outside?"

"Sorry. I didn't know."

Liko stepped out the entrance door and stood in front of the circular driveway, waiting. He suddenly had déjà vu: this was the exact spot where Marimichael had given him a quick hug and kiss, handed him a thumb drive, and then jumped into the back of an SUV.

"Fuck!"

He looked around quickly and was glad that no one had heard his crude outburst. He called Jack and told him about the article that Danielle was working on.

Jack was furious. "Why didn't you say something?"

"Well I'm telling you know," Liko said. And he did, he told Jack everything that Danielle had told him. Even his visit to Ace's farm, the treehouse photograph, and the pending publication in *The New Yorker* magazine.

"Jack, what do you think is going on?"

"I don't know," Jack answered. "Let me see what I can find out and then I'll call you back."

"Thanks. I'll wait here in case she shows up."

An hour passed. The doorman surprised him and brought him a cup of tea. "Would you like to come back inside, sir?"

"No, I'm waiting for a call ... or my friend. But thank you." He did take the hot tea, though.

Several minutes later, Liko received a call.

"Danielle has been taken into police custody." The tone of Jack's voice was not reassuring.

"Why? What happened?"

"I don't know."

"Was it because of the interview with Marimichael?"

"I don't know."

"Does she need bail? Can I pick her up?"

"I'm afraid not, Liko." Jack cleared his voice as if to say something, but then he didn't.

"Is she OK?"

"I believe so. But there will be no bail."

"Why not?"

"Someone in the Attorney General's office called up a district court judge and requested that she be detained."

"Why? What did she do?"

"I don't know," Jack said.

"Don't they have to charge her or release her?"

"That's the way it usually works."

"So how long can they hold her? Twenty-four hours? Forty-eight?"

"I'm afraid it could be much longer. Her right to *habeas corpus* has been suspended."

"I don't understand, but that doesn't sound right. How can that be?"

"A district court justice suspended Danielle's right to *habeas corpus*."

"Dammit Jack! What's going on?"

"Liko, I suggest you go home and get some sleep. Meet me tomorrow at the Deluxe Café."

The thought of meeting Jack at the Deluxe Café was not reassuring. The last time they had met there, someone had tried to kill Jack with a high-powered sniper rifle. "Are you sure?"

"Yes. I'll see you at 9:30 for breakfast."

"OK," Liko said. "If you're sure?"

"I am," Jack said. "Get some rest."

Liko looked at his watch. It was now 10:30pm. He waved goodnight to the doorman, who was standing just inside the entrance door to the lobby, and then he started his fifteen-minute walk back to his condominium, and Charla.

As he walked, he took out his iPhone. This time he activated Aletheia, a personal software program he had not used in a long, long time. Aletheia had been a special gift from his high school biology teacher.

"Hi, Liko," Aletheia said as she re-activated. "How have you been?"

"Aletheia, I'm sorry but I don't have time to socialize."

"I am sorry to hear that, Liko. Perhaps we can catch up later."

"Yes, Aletheia." Liko hesitated, but then said, "I would like that."

"How can I be of assistance?"

"Can the government suspend *habeas corpus*?"

"Liko, I assumed that you meant the United States government. I learned the following: 'At the request of the United States Justice Department, the US Congress passed a statute, which was signed last week by President Charles Clincher, granting the Attorney General of the United States the power to ask a Chief Justice of any District Court to hold any person indefinitely, without trial, suspending their constitutional right to *habeas corpus* during an emergency declaration.'"

Liko didn't know which was more disturbing: that President Clincher had suspended *habeas corpus*, or that Aletheia said she was learning. The first was frightening. The second was ... frightening.

"Thank you," Liko said.

"May I help you with anything else?"

"Yes. Please review all available information on President Charles Clincher and provide me a psychological profile by ... how soon can you do that Aletheia?"

"It depends on how thorough you would like the psychological profile."

"I need it by 8:45 tomorrow morning."

"I will have the report ready for you by 0845 tomorrow

morning. May I have your permission to access information beyond the internet?"

"Yes, you have my permission."

Liko hesitated, but then added, "Good night Aletheia."

"Good night, Liko."

Beyond the internet? How could it do that?

36

Condo Raided

Charla was waiting up for him, as usual. He felt like she was the one steadying force in his life. While he was walking home, Jack had called her and brought her up to speed.

Liko put water in the teakettle, and as it came to a boil, he answered Charla's anxious questions. When the kettle whistled, he prepared her a cup of peppermint tea, her favorite, and himself a cup of Trader Joe's Red Refresh.

They took their mugs of tea onto the balcony. The night air was cool. Since Liko was dressed in jeans and a short-sleeve shirt, Charla pulled a blanket off her bed and draped it around him. She disappeared into the condo for a moment, and when she returned she had slipped into Liko's flannel robe. He smiled.

As soon as the tea cooled, Liko took two maximum strength Tylenol PMs. He planned on going to bed as soon as he felt sleepy.

They sat beside each other in silence for a while, looking at the buildings across the street, gazing into brightly lit apartments and hotel rooms.

"I wish we had a fireplace," Liko said. "The Club has a nice one in their lobby. I like looking at a fire."

"Fireplaces are dirty and smoky," Charla said.

"Don't you find it amusing that they keep their curtains open at night?" Liko asked.

"No," Charla said. "All they do is watch television or play games on their computers or eat."

"I bet the view from their side is more interesting than ours," Liko joked.

"At least the view into my room is," Charla joked. She pulled her legs further underneath her and gathered Liko's robe around her.

Liko turned around to take in the view of her room, wondering what it looked like from outside, looking in. Maybe she was right, he thought.

He saw something coming straight at his head—a blur—and instinctively jerked his head to the side.

The knife cut the skin on his neck above the carotid artery. Liko grabbed the arm and pulled the man out the door, using the body's thrust and weight to spring up beneath him, raising him up, and letting him go. The man sailed over the balcony.

Liko passed through Charla's bedroom and into the living room as the second man stepped out of Liko's bedroom. He too was dressed in a dark suit and tie. Their bodies clashed in front of the coffee table like two linebackers. Liko twisted loose, pivoted, and drove his elbow into the man's chest. Liko heard a snap as a rib broke. And then Charla brought the teakettle down on the back of the man's head.

Liko rushed to the front door, locked it, and swept through his bedroom, walk-through closet, and master bathroom. No one else was present.

Liko rushed back to the living room and stood over the man on the floor. He was face up and his eyes were rolled back into his head. Blood was soaking into the green pattern of the designer rug.

"We need to leave *now*," he told Charla.

She nodded her understanding. Nevertheless, she ran back into her bedroom and reappeared with a pair of blue jeans and a T-shirt. She walked out of Liko's robe as she crossed the living room to rejoin him, pulling on her T-shirt and then hopping into one leg of her jeans. She balanced herself against his body.

He moved into the kitchen and she moved with him. He pulled a paring knife off the magnet holder on the wall and slid it between his belt and pants. Charla grabbed the large butcher knife. She stayed at his side as he moved through the kitchen, his eyes looking at the appliances on the kitchen island and sweeping across the clutter on the counter.

He pulled the roll of paper towels off the paper towel holder, pitching the roll over the kitchen island. The shaft was stainless steel and the base was a heavy block of marble, both held together with a bolt. The paper towel holder felt like a Stone Age weapon.

Together they stepped to the front door. They looked at each other briefly and then Liko unlocked and pulled open the front door. He stepped into the hall, gripping his marble mallet. He glanced up and down the hallway. No one was present.

He extended his free hand back through the doorway. Charla grasped his hand and they ran together to the red exit sign at the end of hallway and then raced down the stairs.

They exited the building through the loading dock and walked a short distance down the sidewalk. In front of them Liko saw Franz, the FBI man, standing over the body of the man who had flown off the balcony. The body lay on the pavement with one leg beneath its back in an awkward, impossible position. Dead. He had to be dead, Liko thought.

How did Franz get here so fast?

Liko pulled Charla to his side and stepped between two parked cars. He watched Franz stare up at the apartment, six floors above, and then back down at the body. Franz had not seen them.

Can we trust him? There's no way he could've gotten here so fast. He must be involved with the guys who attacked us. Franz?

Holding hands he and Charla crossed the street quickly to the Wink. The hotel doors slid open for them and Liko and Charla walked into the lobby, hand-in-hand.

Liko pulled his wallet out of his back pocket and gave it to Charla. "Please get us a room. I need to stop this bleeding before someone sees it. I'll meet you at the elevators, OK?"

"Got it. I'll get us a room and meet you at the elevator."

Liko stepped away to the men's restroom, just off the lobby of The Wink Hotel.

"Let me see that?" Charla said. Liko was peering out the hotel window looking at his condominium across the street. He let go of the washcloth he was holding against his neck.

"I think you'll be OK," Charla said. "Take off your shirt and I'll wash out the blood before it dries."

Liko unbuttoned and slipped out of his shirt.

"You OK?" he asked Charla.

"They never touched me," she answered.

"See that man down there, standing near the body?"

"Yes."

"He's with the FBI. His name is Franz. He was with Jack and me in the Black Hawk helicopter when we flew to the coal plant."

"He got here fast."

"No. I think he was already here."

They looked at each other. Liko saw concern and worry on Charla's face.

A fire engine with siren blaring arrived and drove to the front side of Liko's condominium building, parking on the street between the Ritz Carlton Hotel and the front entrance of the residences. Less than a minute later, an ambulance arrived and parked behind the fire truck, and then a second ambulance arrived. It continued around the block, though, and the driver parallel parked in the street at the back of the residences, close to where the dead man lay on the sidewalk.

Police cars arrived with sirens blaring and blocked off all three sides of the triangular block on which Liko's condominium building sat: New Hampshire Ave, 22nd St., and M St. Their emergency lights flashed off the glass windows of the surrounding hotels.

"What are we going to do?" Charla said.

"I don't know," Liko answered.

Eventually the firemen exited the building and returned to their fire truck. And then the EMS crew appeared at the loading dock with a stretcher, carrying the man that Liko had injured. A policeman moved his vehicle out of the intersection so the ambulance could leave. As it pulled away the driver turned on the siren. The police car followed close behind the ambulance. The fire truck left.

"There's Jack!" Charla said. He had just appeared on the balcony. They watched him peer over the handrail down at the body. Two police had cordoned off the body and stood nearby, warning gawkers away. The ambulance nearby had turned off its flashing lights and now sat idle.

Liko wondered if it was significant that Jack had just arrived, whereas Franz had been there all along.

"Do you trust Jack?" Liko asked Charla. "Jack and Franz are both FBI men and Franz seems to be involved with our attackers. Jack will either side with Franz, or he'll side with us. So do you trust him?"

"Of course I do!"

"With your life?"

"Yes. I do."

Liko went to the hotel room phone. "I can't remember his number, dammit!" Liko had left his iPhone plugged in on his nightstand next to his bed. And his iWatch, too.

Charla recited Jack's phone number from memory.

Liko smiled as he punched the numbers on the face of the hotel phone. "Thanks."

"Hello?"

"Hi Jack."

"Where are you?"

"Can we trust you, Jack?"

"We? Who is we?"

"Charla and I?"

"Of course you can. Where are you?"

"At the Wink."

"Really?"

"Yes, we can see you right now. You're in my living room messing with Zahi's abacus."

Jack looked towards the wall of windows.

"Are you picking up the little microphone you hid in my condo?"

Jack snatched his hand away from the abacus lamp stand and took a step away from it. "Just doing my job."

"Liko," he said, continuing. "You will not be safe at a hotel."

"I know, that's why I called."

"Really, that's why you called me?" Jack laughed. "To tell me you weren't safe?"

"Yes. I know a safe place, but I need your help."

Jack stared out the window at the hotel. "Liko, I can't imagine any place that would be safe for you."

"Correction," Liko said. "I need a safe place for me AND for Charla."

"She's with you?"

"I said she was."

In a loud voice Charla said, "Hi Jack."

"Hi, beautiful."

"Franz is there," Liko said.

"I saw him," Jack said.

Charla spoke up again. "Jack, the dead man almost fell on him. If you know what I mean."

There was a silence for a moment.

"Jack, we'd be safe at your place, wouldn't we?"

Liko thought he could see Jack smiling, but he was too far away. It felt good, though, just to imagine the smile.

"Of course," Jack said. "I'll pick you up in five minutes, at the main entrance to the Wink."

"One more favor, Jack?" Liko said.

"What?"

"Can you pick up my phone and watch? They're by my bedside."

"Anything else?" Jack asked, with humor in his voice.

"You could grab one of my turtlenecks. They're in my closet, on a shelf above my Canali suit."

37

A Friend's Home

"Did you have to throw him off the balcony?" Jack asked as he threw a package of popcorn into the microwave and punched it on.

"He came at me," Liko said. "I reacted." Changing the subject, he said, "Is that a living room or an arboretum?"

Jack smiled. "It's my personal jungle. You like it?"

"I like it," Liko answered. "The palms remind me of Hawaii."

"I like my animals in a zoo," Jack said, "and my plants in pots."

"You must spend hours watering everything," Liko said.

"You've heard of dog walkers?"

"Yes."

"Well, I have a plant waterer. She comes in three times a week and—"

"No he doesn't!" Charla said. "He's teasing you, Liko. But he does shine the leaves."

"Get out!" Liko said.

"He does!" Charla said. "I've witnessed it!"

"One leaf at a time?" Liko asked.

"Yeah," Charla said, smiling at Jack. "He uses vegetable oil."

"No I don't," Jack said, one eyebrow raised. "Keep this up and I won't share my popcorn."

Once the kernels stopped popping, Jack emptied the bag into two bowls. "Everyone want butter and salt?"

"Plain please," Charla said.

"I'll take mine with the works," Liko laughed.

"Really?" Jack melted a quarter stick of butter in the microwave and poured it over one of the bowls of popcorn followed by a generous shake of salt and nori. He added a package of mochi crunch and sprinkled sesame seeds. He tossed everything together. He then handed the bowl to Liko. "For you and me, Liko." He handed the bowl of plain popcorn to Charla.

They followed Jack into his living room and took seats in the jungle, facing a large, wall-mounted television set. A Sonos home sound system surrounded them.

"What should we watch?" Jack asked.

"Something relaxing," Charla said.

"*The Birdcage?*" Jack suggested.

"What's that?" Liko asked.

"A comedy with Robin Williams," Charla said. "I haven't seen it in years."

"It's good fun," Jack said. "You'll like it, Liko."

Liko looked at Charla and she nodded for him to give it a try.

As soon as the movie started, Jack got up and said, "I'll

be right back." He returned with a small bong and a box. "Shall we?"

He set the bong on the coffee table and dimmed the lights. He then sat down and opened the small box and took out a large bud. "An Indica strain. It should help us relax."

Jack had wrapped himself in a stunning manton, a brick red flamenco shawl with yellow hand embroidery and a knotted fringe. His slippers were matching red flats.

Charla ran her hand along his sleeve as he packed the bud into the bong. "It's so soft."

"Spanish silk." Jack flicked the lighter. He picked up the bong, covered the mouthpiece, and set the yellow flame against the bud. He inhaled slowly, lighting the marijuana and pulling the first draft of smoke through the water. The water bubbled.

Jack blew out a stream of smoke. "Relax. This is the last place anyone would look for you." He laughed. "Even me."

Halfway through the movie, Liko suddenly remembered his request from Aletheia. He quickly opened his iPhone and searched for her report.

Jack paused the movie and said, "It looks like this is a good time for a break. Liko, if you need the bathroom, it's through the door, and then down the hall to your right."

Charla got up and went to the bathroom and Jack took the empty bowls into the kitchen.

When he returned, Liko said, "I just found it. A psychological summary of President Clincher. Would you like me to read it?"

"Should we wait for Charla?"

"We could, but when she goes to the bathroom," Liko joked, "she disappears."

"Go ahead, read it."

Liko opened Aletheia's report and read out loud: "Psychological Summary of President Charles Chadworth Clincher. 'Many great and good men sufficiently qualified for any task they should undertake, may ever be found, whose ambition would aspire to nothing beyond a seat in Congress, a gubernatorial or a presidential chair; *but such belong not to the family of the lion, or the tribe of the eagle.* What! think you these places would satisfy an Alexander, a Caesar, or a Napoleon?—Never! Towering genius disdains a beaten path. It seeks regions hitherto unexplored.' Abraham Lincoln, Lyceum Address, January 27, 1838." Liko's expression registered his confusion.

"Aletheia?"

"Yes, Liko?"

"This is a quote by Abraham Lincoln, correct?"

"Yes, Liko."

"Why didn't you give me the summary I asked for?"

"Liko, I can't summarize President Clincher's personality better than this quote from Abraham Lincoln."

"I don't understand."

"President Clincher is of the family of the lion or the tribe of the eagle."

Liko was puzzled. This was the first time Aletheia had failed him. "I don't understand, Aletheia."

"President Clincher is like 'an Alexander, a Caesar, or a Napoleon.'"

Liko looked at Jack and shrugged his shoulders.

"I think I understand what your software program—you call it Aletheia?"

"Yes."

"Liko, I think I understand what it is trying to tell us, and if I am correct, we are in serious trouble. American democracy is in serious trouble. I believe the point is less about genius and more about unchecked ambition."

38

Get Out of Jail Free Card

Attorney Strauss was seated in the attorney-client meeting room at his law firm in Washington, DC. The only furnishings were a rectangular white table and a set of six black plastic chairs. The view out the window was Union Station. The United States Capitol was three blocks away.

Liko was present via Zoom, a face-to-face conferencing program that had become extremely popular during the coronavirus pandemic. His image appeared on the television screen mounted on the wall of the conference room. Liko was seated miles away in Jack's study, with a light blue bedsheet hung behind him from the ceiling to the floor. He was dressed casually in jeans and a harlequin turtleneck. The diamond pattern stretched across his chest in three rows: the top row red, the middle row yellow

254

and the bottom row navy blue, all on a white background. It was his one colorful turtleneck, and, of course, it was the one that Jack had selected from Liko's wardrobe the night of the assault.

Liko had hired Strauss several years earlier to defend a friend, Hugo Haarun, who had been accused of vandalism and graffiti on Maritauqua Island. Strauss' performance had impressed Liko. Strauss also had excellent credentials. He had graduated from Georgetown Law School, JD, *cum laude*, and had been admitted to practice law in the District of Columbia.

"Let me do the talking," Strauss told Liko. "Anything you say, keep it short and to the point."

"Sure."

Strauss's secretary showed a well-dressed man into the room.

Liko could see him via the Zoom connection. Liko said, "This must be him."

"Yes. That's Markowski, the FBI's attorney."

"Who is that with him?" Liko asked.

The uninvited guest was also well dressed. When he removed his FBI cap, Liko instantly recognized him. It was Franz.

"Hello, Mr. Strauss," the FBI attorney said.

"Mr. Markowski," Strauss said. "I appreciate you accepting the meeting. I see you brought someone with you."

"I'm Franz Sauer, with the FBI, working on behalf of the White House. I'm with the team investigating the eco-terrorists. When I heard that Mr. Koholua was going to make a statement, I thought I should be here."

A shiver went up and down Liko's back as he remembered seeing Franz standing over the body of a dead man outside Liko's condominium building, and looking up at Liko's balcony.

"I hope you have no objection to him joining us," Markowski said.

"White House?" Strauss asked.

"That's correct," Franz said.

Isn't there supposed to be a separation between the White House and the FBI? Liko thought. *Guess that ended with the Trump administration.*

"Very well," Strauss said. "Please, both of you have a seat."

Jack then appeared in the doorway.

"I hope that YOU won't mind," Strauss said to Markowski, "but I invited someone from the FBI Public Corruption Investigation Office to join us, Jack Amaya."

Jack entered and stood toe to toe with Franz in a face-off: the bad FBI and the good FBI; the White House FBI and the independent FBI.

"I don't think that's—"

Strauss cut Markowski off. "Before you object, I want to make a confession. Mr. Koholua will not be making a statement."

"Where is your client?" Markowski asked.

"He's joining us via Zoom," Strauss said.

Markowski and Franz turned and looked at Liko's Zoom feed on the wall-mounted television screen behind them. Liko grinned and waved.

"We had to fight DC traffic, dammit!" Markowski said.

"You should have taken the metro," Liko said, joking. "Union Station is just a block away."

Liko saw Strauss give him a quick look that said "be quiet."

"Then why did you ask for this meeting?" Markowski asked Strauss.

"Mr. Markowski, last week I reached out to you on behalf of my client and you made us an offer. You said that your office would consider a life sentence and not the death penalty if Mr. Koholua would turn himself in, confess, and plead guilty to the murder of a federal law enforcement official. Well, today he'd like to make a counteroffer."

"This is bullshit," Franz said, looking at Liko on the monitor. His tone was both condescending and angry. "You're wasting our time."

"If I were you," Liko said, "I'd listen to what my attorney has to say."

"Fuck you," Franz said. He pushed his chair away from the table and stood up.

Unfazed, Strauss said, "My client, Mr. Liko Koholua, will work with the federal prosecutor—and agree not to sue him and the US Attorney General's Office—provided two conditions are met."

"Don't waste my time." The White House's FBI man opened the door and stood in the doorway. "Coming?" he asked Markowski.

"No," Markowski said. "I'll hear what they have to say. Not because I think it will be productive, but because I'm in no rush to get back into that traffic."

"Suit yourself," Franz pulled the door closed behind himself.

"As I was saying," Strauss said, "before someone became emotional and stormed out—"

"Two things, you said?"

"Yes. First, you must release Danielle Queen." Strauss paused and added, "I believe she is being held without being charged with a crime and without bail, and for more than twenty-four hours?"

"No comment."

"For your sake I hope that she is not." Strauss looked intently into Markowski's eyes. "Second, I want any and all charges against my client, Liko Koholua, dropped."

"Please tell me," Markowski said, smiling, "why the government would agree to such terms?"

"My client has information that would be of special interest to everyone involved in this case."

"Really?" Markowski chuckled. "I apologize for laughing, but really? What planet are you from?"

Strauss opened an envelope and handed Markowski a photo. He allowed him a moment to look at it and then said, "That's a photo of Ms. Marimichael O'Brien's leadership graduating class. It was taken at the National Defense University on Fort Leslie McNair, in DC."

"Yes, I recognize her. What is the significance?"

"Do you recognize the general? Standing to the side of the class?"

"President Charles Clincher?"

"That is correct. He also taught one day during her leadership training."

"I'm sure the general had his picture taken with lots of people."

Strauss handed Markowski a second photo. "This is a picture taken many years ago. Do you recognize anyone?"

Markowski studied the photo for a moment. "The terrorists Marimichael O'Brien and her two brothers, Shane and Fintan."

"And the other boy in the photo? Do you recognize him?"

"No."

"The one with the chubby face and wearing glasses. At the foot of the treehouse. Looking up at Marimichael. With admiration on his face. Perhaps even awe?"

"No—" the color drained from Markowski face.

"It's the President of the United States," Liko said. "He spent childhood summers visiting his grandparents, who lived near Marimichael's parents."

The attorney now shared other photographs that Ace had taken of Charles and Marimichael together. Ace had emailed the additional photos to Liko after Ace learned that Marimichael was dead, Danielle had been detained without *habeas corpus*, and that Liko's life was in danger.

"Charles and Marimichael had an ongoing relationship," attorney Strauss said. "I imagine the public would be very interested. Don't you?"

Markowski swallowed and recomposed himself. "Just a moment." He punched a number on his phone. "Franz, get your ass back here," he said. "We have a problem."

Markowski turned his attention back to Strauss. "What do you want?"

"As I said, my client has two requests: the immediate release of Ms Queen, and that all pending charges against him be dropped." Strauss opened his hands and spread them apart. "Do we have a deal?"

The White House's FBI man, Franz Sauer, opened the door. He turned to Markowski and asked, "Is there a problem here?"

"No problem." Jack clapped his hands rhythmically in crisp, loud, and fast fuertas, followed by a loud slap to his thigh. "Liko just played a 'get out of jail free card.'"

Franz's smile faded.

Huh-huh-huh, Liko thought, barely keeping the noxious laugh to himself.

Danielle was furious and wanted no part of the deal. She insisted that she was not bound by the agreement reached between Strauss and Markowski. She angrily asked Strauss, "How could you presume to represent me? We have never even met."

"Because we had no access to you," Strauss said. "You were being held without bail, indefinitely, at an unknown location, and your *habeas corpus* rights had been set aside by a district court judge, one of the many unqualified judges appointed by the Trump administration. What other choice did we have?"

"It's still wrong," she said. She scolded Liko, too: "You have compromised my integrity as a journalist! How dare you!"

As soon as she returned home, Danielle emailed the first story of her mini-series to *The New Yorker*, along with her signature on documents for its immediate publication. She included the clubhouse photograph along with the new pictures that Ace had given Liko. She also sent them to Sid, with a request for him to validate them before publication.

We're fucked, Liko thought. *As soon as Danielle's story is published, we're all fucked.*

"The O'Briens and the President of the United States, the Untold Story," appeared in the next weekly issue of *The New Yorker*. The magazine made Danielle's article and the ongoing series free, just as it had made its coronavirus news coverage and analysis free for all readers during the pandemic. Danielle's article immediately went viral, worldwide.

39

United States Climate Force

"Do not fear, I am in control here." President Clincher waved at the members of the United States Congress. He had used his authority under Article Two of the United States Constitution to convene both houses. "The rumors of a *coup d'état* and my demise are exaggerated."

The members of Congress clapped and cheered, "Long live Clincher!"

Liko, Charla, and Jack were seated in Jack's jungle watching CNN's live coverage. Tonight there was no popcorn and no one was smoking.

"Today I asked for the resignations of the Joint Chiefs of Staff." President Clincher waved several sheets of paper, crumpled them slightly in his hand, and set them on the podium.

"It's like the political thriller *Seven Days in May*," Jack said, "but in this case, the president led the coup and the president took total control of the military. In the movie, it was just the opposite: the military tried to take over the presidency."

"But Clincher succeeded," Liko said, "whereas the military in the movie failed."

"It appears so," Jack said.

President Clincher, as usual, cut to the chase. "Today I have another reason for all Americans to rejoice. Today I am asking Congress to pass Senate 1215: the United States Climate Force Authorization Amendment. I am announcing today that the Department of the Climate will be a military department within the Department of Defense, joining the Department of the Army, Department of the Navy, and Department of the Air Force.

"Senate 1215 amends this year's National Defense Authorization Act and provides $150 billion dollars for military activities related to the new Department of the Climate."

Audience members broke into applause.

"The mission of the Department of the Climate is to organize, train, and equip climate forces in order to protect US and allied interests, worldwide, and to provide climate capabilities to the United States Joint Force."

Democrats jumped to their feet to applaud. Republicans (even those who had funded and created Trump's United States Space Force) looked unhappy and sat glued to their chairs.

"The Department of Climate's major responsibilities are: First, prevention of human and animal extinction. Second, offensive and defensive climate control. Third, to work closely with commercial leaders in the climate domain."

The audience cheered and Republicans glared.

"Tonight I am unveiling the Climate Force Shield, which will soon be worn proudly by all military climate personnel in the new department. As you can see, the shield is an owl perched on an hourglass. The hourglass is almost empty." Liko could not believe his eyes, and wondered if anyone else made the connection with the terrorists' use of the owl symbol.

"If you look closely, you will see the owl's rounded head, short tufted ears, alert eyes, sharp beak, and merciless talons. If you look even closer, you will see a small green and blue planet, in an elliptical orbit, circling the hourglass."

"I need a gin and tonic," Jack said. "Anyone else?"

"Let's get the bong out," Charla said. "Shall we?"

Liko said, "I don't know if I should cheer or laugh or cry."

"Yeah," Jack said. "Let's get stoned." He left the jungle to get the bong and his box of buds.

Liko thought about everything that had happened since he moved into Zahi's condo. He thought about the first time he saw Marimichael at the Dupont Farmers Market. He thought about dinner with her brother Shane, Danielle's presentation at the Cosmos Club, Jack's first visit to ask for Liko's help; he thought about all the assassinations, murders, protests, violence, and civil disobedience; and then he remembered Marimichael's swan dive.

If the president's men want to find me, they will. There is nothing I can do. I'm at the mercy of the most powerful man in the world.

"It is what it is," Liko said. "But I wouldn't have done anything different. How about you Jack?"

Jack had resigned from the FBI, taking an early retirement. He had told them that he intended to devote himself to dancing—that tango and flamenco were the only things that might keep him sane and balanced.

"Let's get stoned and then, if you want, we can talk about it." He smiled at Liko, took a hit on the bong, and passed it to Charla.

It troubled Liko deeply that Charla was probably in danger. But what could he now do? "I'm sorry I got you involved in all this, Charla."

"Liko, it's not your fault," she said. "You know I love the drama. I wouldn't have it any other way, either." She took a slow, long, deep hit on the bong and exhaled, then passed the bong to Liko.

He held it in his hands. "It is what it is," he repeated. He inhaled and exhaled the smoke. It was moist and cool. Jack had placed some ice cubes in the water.

From now on, Liko thought, *I'll take my bong with ice.* He smiled.

The buzz hit him and the first thing he thought was *I need to call a lawyer and make a will.* He wondered if Jack and Charla had wills. He decided it wasn't the right time to ask them.

"I wonder how Hugo is doing," Liko said to Charla.

"Who is Hugo?" Jack asked.

"Pass me the bong again and I'll tell you," Liko said. "It's a long story."

And then he thought about Danielle again. It saddened him that Danielle would not return his calls. He had tried email, text message, and even snail-mailed an 'I'm sorry' Hallmark card, but no answer.

He knew it was his fault. His poor judgment. What a stupid idea to use Danielle's article to negotiate her release from the detention center!

I didn't realize it at the time Danielle, but I betrayed you. I didn't understand how much freedom of speech meant to you. He remembered that Hugo was like that too, with his painting. *I would never willingly compromise my core values, so why did I think that you would compromise yours? I should have known that.*

Charla handed Liko the bong. It was his turn again.

"Can we play some music?" Liko asked Charla and Jack.

"Great idea," Charla said.

"What would you like to hear?" Jack asked.

"A rapper and environmental activist named Xiuhtezcatl."

Jack raised an eyebrow. "Siri's connected to Sonos."

"Got it," Liko said. "Siri, play Xiuhtezcatl Tonatiuh Martinez' 'Broken.'"

"What?" Charla said.

"Try these?" Jack handed Liko a pair of Sony noise-reduction headphones.

"Thanks." Liko smiled sheepishly.

Jack walked behind him and massaged his shoulders until he relaxed. Jack then sat down next to Charla and put his arm around her.

When "Broken" ended, Liko said: "Siri, play Paul McCartney 'Despite Repeated Warnings.' And then, Siri, cue up songs about climate change."

Liko closed his eyes and imagined.

The next morning, all the major papers reported the highlights of President Clincher's speech. Over the following few days, world leaders were quick to respond.

Beijing verbally attacked President Clincher for weaponizing the climate crisis. Russia announced it would double oil production. And the people of Brazil awoke to discover the US Navy afloat off their coast.

The Secretary General of the United Nations called for a special meeting of the UN climate change conference in New York. Clincher recalled the existing US Ambassador to the United Nations and appointed Brigadier General Jones, who had facilitated the Department of Defense Climate Change Adaptation Working Group. The President then announced that both he and Brigadier General Jones would attend the upcoming UN conference.

Clincher asked all current US ambassadors to tender their resignations. He reappointed a few, but he replaced most with pro-climate business leaders who supported him and his administration's aggressive climate agenda.

Each morning, Liko looked forward to reading the latest climate news. Each evening he expected to die.

40

The Great
Horned Owl

A voice said, "Mr. Koholua, a man from the FBI is here to see you. Should I send him up?"

Déjà vu. Liko knew that this FBI man was not Jack. If it was Franz Sauer, then the White House's FBI man had better have a parachute. "Send him up," Liko said.

Liko had stopped by his condo to pick up a few things for himself and Charla. He had filled two Trader Joe's shopping bags with the stuff Charla said she couldn't live without, including silk underwear, a cosmetic bag, and, of course, her kimono nightgown. He hadn't even started to pack any of his own stuff.

Of course. They were watching. They were waiting for me to return.

He glanced around the condominium, scanning for anything he could use as a weapon. He still had not replaced his paper towel holder. Knives were lined up on the magnetic strip, but he shook his head, no.

His eyes settled on Zahi's abacus lamp stand. He laughed nervously. *Zahi, I'm not going to ruin your lamp.*

He stepped to the front door and took a deep breath. He then unlocked the door and opened it wide.

The FBI man had just arrived. The unexpected opening of the door caught him off guard and he almost fell into the condo. He regained his balance.

He was dressed in a dark suit. He was holding an FBI badge and ID in his hand.

Holding the door open, Liko asked, "Are you coming in? Or do I go with you?"

The man appeared taken aback at Liko's nonchalance. Suspicious. He closed his badge and slipped it into his back pants pocket. "Come with me."

The walk down the hall, the elevator ride, and the walk through the lobby were intense. Liko thought the man appeared nervous. Liko felt confident that he could subdue the man if he had to.

The front door man said, "Good evening, Mr. Koholua."

"Good evening," Liko responded.

"Would you like an umbrella?"

"Is it going to rain?"

"The weather looks threatening."

Liko understood exactly what the doorman was telling him. "No. But thank you."

The black Cadillac Escalade SUV was waiting at the curb. As he was led to the car, Liko could see the driver inside. Another Mormon wannabe.

When the FBI man opened the back door of the large SUV for him, Liko was surprised to see Danielle seated behind the driver. Liko slid into the back seat and the FBI man closed the door and slipped into the front passenger seat.

Liko looked at Danielle and quipped, "I hope this is your Uber."

She looked at him and smiled at his joke. Then she turned her head and looked out her window, on her side of the car. She held her hand out, though, across the middle of the seat. Liko placed his hand in hers. He looked out his window and sighed. He felt her squeeze his hand. They turned and looked at each other and Liko mouthed the word "Shit!"

Danielle and Liko were seated across from the President of the United States in the Oval Office.

"I read your blog about Marimichael and her rebels, the men and women she recruited. Well done! And your *New Yorker* series on Marimichael was excellent! I'm sure she would have been pleased. I am." With a smile, the president said, "I like your work."

Danielle did not respond. Liko guessed that she was like him, still in a state of shock.

"And Mr. Koholua," the president said to Liko, "Marimichael spoke well of you. She was impressed with your drone work at Davos. She had a passion for birds and butterflies and drones. Anything that flew."

How would he know that? Liko thought. He nodded.

"The attack on you in your home was unfortunate."

Unfortunate? They wanted to kill me!

"My FBI is sometimes overzealous in protecting me."

My FBI?

"Why have you brought us here?" Danielle asked.

"I would like you to write my biography."

"A biography?" She took a breath and said, "About you?"

"Yes. I want you to be my official biographer." After a beat, he added, "I expect several volumes."

What a fucking narcissist, Liko thought.

"Why me?"

"Because you have shown courage." Clincher looked at her. "You're not afraid of me, are you?"

"Of course I am," she barked, unable to control a loud, explosive laugh.

Clincher turned to Liko. "You see, she just proved my point. Not only does she write the truth, but she also speaks honestly. She doesn't varnish anything. Do you know how rare that is?"

"No," Liko said. "But I do know that she will not compromise her values." *Why did I say that? What a stupid thing to say!*

Clincher turned back to Danielle. "That is why I have chosen you to be my biographer."

"Did you hack into the Delaware Law Office servers?"

"Yes."

"Did you assassinate all those people? President Danson? How is that not treason? The CEOs of fossil fuel companies? The oil barons? The politicians? The Supreme Court Justice?"

"Not personally."

"Did you plan and lead the recent military coup?"

"Yes."

"And you want all this in your biography? Your official biography? Because I can write it no other way."

"I did what had to be done. Nothing more, nothing less. I want you to shine a light on everything that has happened. It has all been necessary."

"Necessary?"

"It was necessary to prevent the sixth extinction of humans and other animals. I have done nothing but what has been necessary. I do not shrink or hide from what I have done, just as you do not shrink or hide from writing the truth as you understand it. You shine a light on everything."

The president continued: "I want you to be truthful. I want you to tell the good, the bad, and the ugly. Yes, the good, the bad, and the necessary. It will be the story of how I led the United States, and the world, to prevent the sixth extinction of humans and other animals. It will be my story."

Chills ran down Liko's back.

A woman appeared at the door to the Oval Office. "An urgent message for you, sir." The president motioned for her to enter. She walked across the room and handed him a notecard.

He took a few seconds to read the card. He smiled and said, "Not that urgent." He folded the message and put it in his shirt pocket.

"I have just one more question," Danielle said to the president. "On the record?"

"Yes, but then I must return to work. Fire away."

"In the clubhouse photo, you are holding an animal by the tail?" Danielle asked. "Is it a decapitated cat?"

"Yes. The farm cat Topaz." President Clincher smiled. "He was my gift to Marimichael."

"But why?"

"He was killing Marimichael's martins."

"Mr. Cody Long Photoshopped you out of the clubhouse picture. Did you ask him to do that?"

"Heaven's no."

"So why did he do it? Why did Mr. Long Photoshop himself into the picture and you out?"

"I believe he was jealous," Clincher said with a slight smile.

"Jealous? Of what?"

"Danielle, I thought you knew," he said. "We were all in love with Marimichael."

"I have a question, too," Liko said. "If I may."

Danielle and President Clincher turned their attention to Liko. Liko thought he detected surprise on Danielle's face, but curiosity on the president's.

"What would you like to know?" the president asked Liko.

"You said that you decapitated the cat?"

"Yes, I did."

"The Great Horned Owl decapitates its prey, too. I noticed that the rebels used the owl as their symbol. Any comment about that?"

President Clincher stood up from his chair. His movement was so abrupt that Danielle and Liko stood up too.

"Please sit down," he said, motioning for them to sit down and they did.

Clincher walked to the Seal of the President of the United States that was displayed on the floor of the Oval Office. He stood on the bald eagle, one foot on the olive branch and the other foot on the bundle of thirteen arrows, both held in the eagle's talons. The background of the shield was azure and the perimeter yellow.

He turned around to face Liko and Danielle. Behind the president was his desk and the American flag, which stood at attention in front of gold curtains. The curtains had been pulled open to allow sunlight to enter, and it shone on the President of the United States.

Liko slid to the edge of his seat. As he did, Liko noted that a guard tensed, leaned in.

The president took off his dress jacket. He held it to his side and the guard stepped forward, immediately, and took it from him.

He's placed himself between me and the president. Liko admired the guard for that. He was large and powerful.

The president undid his red tie, pulling it from around his neck. He handed it to the guard. Next, he unbuttoned his white shirt, pulled it out from his trousers, and took it off, handing it to the guard. White hair peeked over the edge of his V-necked undershirt.

The guard stepped to the side, yet kept himself in an advantageous position relative to Liko and the president. *He'd die for the president,* Liko thought. *I couldn't take him.*

The president turned around so that his back faced Liko and Danielle. As President Charles Clincher pulled off his undershirt in the Oval Office, Liko watched, his mouth agape, like a fledgling.

A Great Horned Owl gazed out from the back of the President of the United States, the yellow eyes fixed on Liko and Danielle. Liko admired the rounded head; the prominent, tousled, ear-like tufts of feathers; the short, sharp bill. The owl held an hourglass in one of his curved, pointed talons. In his other talon he held the Earth. It was similar to the Climate Force Shield that soldiers in the newly created Department of Climate would soon be proudly wearing.

I wish I had my iPhone. Security had checked all their electronics when they entered the White House. Hopefully they would be returned when they left.

The White House photographer stepped forward and took a series of quick photographs.

The guard never took his eyes off Liko.

That evening President Clincher announced the creation of a federal news agency. "Papers are no longer commercially viable," he said, "but they have a social value that must not be lost. That is why I have requested Congress to create a federal news agency."

"That should kill the last of the local weeklies," Danielle said. Her voice was filled with sarcasm. After the meeting with the president, she and Liko had made up. She was now seated with him, Charla, and Jack in what Liko called Jack's Jungle Man Cave. "It's the death of the afternoon papers, too," she added.

"I have asked the Speaker of the House to fast track the legislation and send it to my desk as quickly as possible. Our goal is to have the new agency up and running in the next few months. When Congress sends me the legislation, I will sign it."

"Well, fuck!" Danielle said. "What we need is for the federal government to provide non-revocable grants to local communities so they can fund local newspapers and digital news. A federal news agency is not what we need!"

41

Cemetery Walk

Charla had given up hope that Liko would ever go through Zahi's personal belongings, so she did a spring-cleaning herself. She either threw out Zahi's possessions or donated them to charity. She dropped off his suits, dress shoes, and jackets at the Goodwill on Glebe Road. She donated boxes of his hardcover books, mostly about economics, to the West End library.

She set aside a few personal items for Liko, including a Nikon mirrorless camera—a Z6. She found the camera and a prime lens in a shoebox along with the original owner's manual. She charged the lithium battery overnight and set the camera on the coffee table. She then discovered that the camera was missing a memory card to store photos, so she walked to a local camera shop and bought a Sony XQD 64GB card. She complained to the store manager that it was ridiculously expensive—$130—so when she set it next to the camera, she included the sales receipt so Liko would appreciate its cost.

Jack had wholeheartedly endorsed Charla's idea of encouraging Liko to take up photography, and had gifted him Joel Sartore's National Geographic book, *The Photo Ark Vanishing: The World's Most Vulnerable Animals.* At Jack's request, Charla had set his gift, wrapped in green tissue paper, on the coffee table next to the Nikon. "Thank you for the memory card," he told her. He wished that he could thank Zahi, too.

Liko couldn't believe the thoughtfulness of their gifts. He sat on the couch with Charla, looking at the amazing photos of soon-to-be extinct animals. Liko picked up the camera and took his first picture—Charla, crying, holding Sartori's book in her lap, a tortoise shell peineta in her hair. Liko never set the camera down.

He fell in love with the Nikon Z6.

Jack broke his stride. They were walking among the gravestones in a small rural cemetery several miles down the road from Marimichael's childhood home in West Virginia.

Liko paused to take a photo of three upright gravestones on the top of a hillside directly ahead of them. It was a cloudy day, which he now knew was good for photography. *I don't have to worry about shadows,* he thought, *and the white balance is good. Even for new, white grave markers.*

Liko snapped several pictures using a prime lens.

"Remember my friend Hugo, the artist? I told you about him?" Liko told Jack. He turned the camera vertically and took another shot. "He lost his right hand and several fingers on his left hand, but it didn't stop him from creating. He exchanged his brushes for cans of spray paint. Photos of his art continue to go viral. Even several photos of his art that I took with my iPhone."

"Painting and photography are both powerful tools," Jack said.

More powerful than my fists. "It's an especially powerful tool in the capitol." Liko grinned and added, "Lots of protests in the District."

"Protesting and photography are both protected by the Constitution," Jack said. "Or at least they used to be."

"You're not going to start lecturing us about the Constitution again, are you?" Liko chided. He glanced at Charla and Danielle. They both returned a look that said *We hope not, too.*

"No," Jack said. "Not this afternoon."

"When does your dance studio open?" Liko asked.

"The grand opening will be next week, Monday," Jack said. "I hope you can make it."

"I wouldn't miss it," Liko said.

"And Liko's going to enroll in your ballroom dance class," Charla said. "Waltz, foxtrot, bolero, tango. Right Liko?"

Liko looked like a deer caught in the headlights of a car. He couldn't imagine taking dance lessons, but then he couldn't imagine disappointing Charla, either. Or Jack. "I'll be there," he said, "to support you in your new business."

They stopped in front of the three graves. In the center was Marimichael. She had been cremated and interned between her two brothers, Shane and Fintan.

"Here we are, out in the middle of nowhere, and the graves have already been vandalized." A swastika and profanity had been painted in red and black spray paint.

Liko took several pictures of the gravestones, capturing the vandalism. He uploaded the best photos to his new Instagram site.

"I imagine Marimichael would be happy with the way things are going. The Climate Defense Tax," Liko didn't call it the Paris Hilton Climate Tax like the administration, "the United States Climate Force, and now the war on Brazil."

Just last week President Clincher had asked the United States Congress to declare war on the Federative Republic of Brazil. "We will no longer sit idly by while Brazil destroys the earth's greatest ecological resource, the Amazon rainforest," the president had told the nation. "Amazonia does not belong to one nation but to the entire world. To the leaders of Brazil, Americans say: Enough is enough. Stop destroying the Amazon."

"I'm sure she would be happy," Jack said, agreeing with Liko.

"She'd be on the front lines," Charla said, "parachuting into the jungle."

"First Brazil, then what?" Liko asked.

"Who knows," Jack said.

"At least Clincher's candid," Charla said. "No bullshit."

"Yeah," Liko said. "Not like the last few wars."

"What do you mean?" Jack asked.

"No weapons of mass destruction? No bogus provocations like the sinking of the USS Maine or the Tonkin Gulf incident? No fabrications like Nayirah's testimony before Congress about Iraqi baby killers in Kuwait?" Liko smiled sadly. "You're right, Charla. At least now there are no lies, no deception. We are going to war with Brazil to save the Amazon."

"If he wasn't a despot, there'd be no downside," Charla said.

Her statement was so sad that all three of them had to shake their heads.

"He's forthright, yet ruthless," Liko said.

After a few moments of silence, Jack said, "So what are you going to do now, Liko?"

"I'm going to cover the elections this fall," Liko said. "Maybe there will be some protests."

Liko smiled at Danielle. She had fallen quiet. "Danielle, you once told me, 'Our power is what we shine the spotlight on.' When you start working on your next project, let me know. Maybe I can take some pictures for you."

"I will," Danielle said.

"Do you wish you had taken him up on his offer?" Liko asked Danielle. "To write his biography?"

"Hell no," she answered without any hesitation.

"Why not?" Liko asked.

"Sooner or later, like all despots, he would want to make revisions, he would want to rewrite history—or at least try. And then I would have been in his way."

"I don't expect he'll live long," Jack said.

"Why do you say that?" Danielle asked.

"Most revolutionaries don't. History teaches us that someone always rises up to derail things. To seize power for themselves."

"Have you heard something?" Danielle asked Jack, her voice anxious.

"No. But I don't expect Clincher will die peacefully in his sleep." Jack turned to Charla with a smile and a bow, "Dance with me?"

"In the cemetery?" she replied.

"I don't know why not." Jack took Charla in his arms, and they walked to a grassy area on the other side of Marimichael and her brothers' gravestones.

"Jack!" she exclaimed as he led her into an underarm turn.

Liko set his camera on Marimichael's headstone and extended his arms to Danielle. "May I have this dance?"

"Yes," she answered.

"I'm not very good," he warned. He followed Jack and Charla into the open grassy area.

"Let me lead," Danielle suggested.

"My Uncle Keahi, he danced the hula. I wonder if he ever tried ballroom dancing?"

As they danced, the tension, the burden of the times, fell from Liko's shoulders. His moves were clumsy, as always, but it didn't matter.

The following week, President Charles Chadworth Clincher stood at the podium after being introduced by the Secretary General of the United Nations. After the applause died down, he cleared his throat and said, "Thank you very much, Secretary General. Mrs. Secretary-General, distinguished delegates, ambassadors, and world leaders: As I address this hall as President of the United States for the first time, let me welcome all of you to this special meeting of the United Nations General Assembly in New York. My message today will be brief and to the point:

"The United States will not rejoin the Paris Agreement, which is a voluntary, uncoordinated agreement burdened not only by freeloading but also by short-sighted, venal leaders.

"The United States will not attend yet another

Conference of Parties, COP*whatever*, to *discuss* the climate crisis. Such non-productive meetings have been held for more than forty years and they have accomplished nothing. We are no closer to reining in the climate crisis today than when some of your fathers attended COP1. Enough is enough. We will no longer repeat the failures of the past.

"Today I am announcing a new organization: the Conference of Parties Climate Club. I am inviting all of you to join the United States of America and become charter members. Membership will be costly, as it is to join any worthwhile campaign.

"Those nations that decide not to join will quickly learn that the cost of non-participation will be much greater than the cost of membership. Members will impose heavy tariffs—*very* heavy tariffs—on all imported goods from those countries that foolishly decide not to participate. And, *as necessary*, the full force of the United States military will stand beside all members who agree to, and who play by, the rules of the club.

"What are the rules? There are only a few. First, members shall agree to long-term, worldwide goals for reducing carbon emissions. Second, members shall agree, and adhere to, a minimum domestic carbon price. Three, members shall agree to enforce penalties against those who fail to participate.

"I want to be perfectly clear: from this day forward, there will be no more free rides! We are face to face with an invisible enemy, greenhouse gases. We are in a worldwide climate crisis. The COP Climate Club will behave accordingly.

"I now want to introduce to you the new United States Ambassador to the United Nations, Brigadier General

Jones. He will explain the logistics of the COP Climate Club, including the time and place of its first meeting. I invite you to join us. Thank you very much."

Lightning Source UK Ltd.
Milton Keynes UK
UKHW041314070620
364543UK00007B/196/J